D1222160

Reiner Braun and David Krieger (Eds.)
Einstein – Peace Now!

Einstein – Peace Now!
Visions and Ideas

Edited by Reiner Braun and David Krieger

WILEY-
VCH

WILEY-VCH Verlag GmbH & Co. KGaA

The Editors of this Book

Reiner Braun
Max-Planck-Institute for the
History of Science

David Krieger
Nuclear Age Peace Foundation
Santa Barbara, USA

1st Edition 2005
1st Reprint 2006

All books published by Wiley-VCH are carefully produced. Nevertheless, authors, editors, and publisher
do not warrant the information contained in these
books, including this book, to be free of errors.
Readers are advised to keep in mind that statements,
data, illustrations, procedural details or other items
may inadvertently be inaccurate.

Library of Congress Card No.: Applied for

British Library Cataloguing-in-Publication Data:
A catalogue record for this book is available from the
British Library.

**Bibliographic information published by
Die Deutsche Bibliothek**
Die Deutsche Bibliothek lists this publication in the
Deutsche Nationalbibliografie; detailed bibliographic
data is available in the Internet at
<http://dnb.ddb.de>.

© 2005 WILEY-VCH Verlag GmbH & Co. KGaA,
Weinheim

Printed in the Federal Republic of Germany

Printed on acid-free paper

Typesetting TypoDesign Hecker GmbH, Leimen
Printing and Binding Ebner & Spiegel GmbH, Ulm
Cover Design Himmelfarb, Eppelheim,
www.himmelfarb.de

ISBN-13 978-3-527-40604-3
ISBN-10 3-527-40604-2

Table of Contents

Einstein – Peace Now! Reiner Braun and David Krieger (Eds.)
Copyright © 2005 WILEY-VCH Verlag GmbH & Co. KGaA, Weinheim
ISBN 3-527-40604-2

Einstein's Importance Today

Reiner Braun and David Krieger

Albert Einstein was a citizen of the world and a deeply committed internationalist. He spoke out regularly on issues of peace, expressing his high ideals for peace and world government. He firmly opposed militarism and narrow nationalism and expressed such views publicly throughout his life. The following statement impressively summarizes his principle pacifism from an article he published in October 1930 in the work "Form and Century" with the title "How I see the world?":

> This brings me to the worst outgrowth of herd life, the military system which I abhor. I feel only contempt for those who can take pleasure marching in rank and file to the strains of a band. Surely, such men were given their great brain by mistake; the spinal cord would have amply sufficed. This shameful stain on civilization should be wiped out as soon as possible. Heroism on command, senseless violence and all the loathsome nonsense that goes by the name of patriotism – how passionately I despise them! How vile and contemptible war seems to me! I would rather be torn limb from limb than take part in such an ugly business. I happen to think highly enough of mankind to believe the specter of war would long since have disappeared had the sound common sense of the people not been systematically corrupted by commercial and political interests operating through the schools and the press. [1]

His entire life he remained true to his principles. As a youth he resisted entering into the hated military service. As a young man, instated only a few months before at the Prussian Academy of Sciences, he condemned the First World War by signing the "Appeal to the Europeans," which attacked Prussian militarism. This plea for peace was signed by two others and was a response to another document, "Appeal to the Cultural World", which supported Germany's war of aggression and had been signed by nearly the entire German intellectual and cultural elite. Worldwide he actively supported conscientious objectors to the war and was an untiring exhorter and ac-

Einstein – Peace Now! Reiner Braun and David Krieger (Eds.)
Copyright © 2005 WILEY-VCH Verlag GmbH & Co. KGaA, Weinheim
ISBN 3-527-40604-2

tivist for disarmament. The aim of his anti-militaristic and pacifistic activity was to move individuals and international institutions to overcome war.

Einstein's stance is that of a "rational pacifist". In addition to a profound moral and ethical objection to war, Einstein's attitude is essentially founded on scientific argumentation and a politically pragmatic approach toward the problem of securing peace. In the final decade of his life, after atomic weapons had been used at Hiroshima and Nagasaki, Einstein devoted his considerable energy to attaining peace in the Nuclear Age.

A few days before his death, with his last act for peace, he signed the Russell–Einstein Manifesto, which continues to be the most famous appeal of its kind and is forever tied to his name. The appeal concluded:

> In view of the fact that in any future world war nuclear weapons will certainly be employed, and that such weapons threaten the continued existence of mankind, we urge the governments of the world to realize, and to acknowledge publicly, that their purpose cannot be furthered by a world war, and we urge them, consequently, to find peaceful means for the settlement of all matters of dispute between them. [2]

Thus it is written in the manifesto, a document characterized by profound humanity and great concern, while conveying hope for a peaceful future for mankind. The manifesto came into being following an intense exchange of correspondence between Albert Einstein and the mathematician and Nobel Prize recipient for Literature Lord Bertrand Russell. In the manifesto, Einstein joins Russell in expressing his impressive humanity with the matter-of-fact and cruel, yet inevitable question: "Shall we put an end to the human race, or shall humankind renounce war?"[1]

When could it make more sense to remember the pacifist Albert Einstein than on the 100th anniversary of his *annus mirabilis*? In 2005, the entire world celebrates the 100th anniversary of Albert Einstein's great scientific discoveries of 1905. UNESCO and many scientific societies have proclaimed 2005 the "International Year of Physics". Germany celebrates the Year of Einstein and many international organizations will remember him.

1) See Russell-Einstein Manifesto, Appendix, pp. 289–292.

A number of institutions boast his name. The politicians celebrate him, or perhaps it is more accurate to say that they attempt to instrumentalize him. Apart from his scientific authority, which is often extolled simply as genius, the social critic, and, indeed, the pacifist is named in festschrifts and commemorative publications and then often dismissed. Everything then quickly returns to "normal" science lacking in serious social reflection in the seeming ivory tower and to militaristic everyday politics – be it the continuation of the wars in Iraq and Afghanistan, be it the militarization of Europe, be it heightened level of armaments in Southeast Asia.

Albert Einstein cannot, however, simply be dismissed or ceremoniously disposed of after the commemorations. His scientific achievements have endured for the past 100 years and continue to act as a motivation and role model for many scientists. His critical remarks on the ethics of science and responsibility of scientists promote reflection worldwide, especially among students, a reflection on the meaning, truthfulness, and goals of science. His endeavors toward disarmament have served for decades as an example for dedicated scientists united in international peace organizations, such as the Pugwash Conferences on Science and World Affairs and the International Network of Engineers and Scientists for Global Responsibility (INES). Moreover, his open-minded thinking still promotes and encourages individuals to stand up for what they believe in. While Albert Einstein did not yet know the term *whistleblowing*, his stance has nourished it. His remarks on the responsibility of the individual scientist are manifold; as too are those on the structures of science that make responsible action possible in the first place. Even today it remains worthwhile to reflect deeply on these remarks.

In a letter to a student, Einstein also expresses his fundamental enthusiasm for science: "That which can be achieved through the sheer delight of recognition is a versatile instrument in the hands of a living human being." [3]

The attempt to comprehend and understand Albert Einstein means to perceive him in his totality in the period of his work and within the society that molded him. This is how we must experience this multifaceted personality with his contradictions as well. He was a gifted scientist, pacifist and open-minded thinker; he was a media star, father, lover and husband, loner, "bachelor", "Einspänner", and a pugnacious debater. In other words, he was a true human being as well as an icon, myth, and genius to the public. Einstein himself spoke of genius in this way:

> I have never understood what 'genius' really means. The term as such does not add to our understanding of art, science or literature. It seems to me that in order to understand a work, our aim must be to understand the tools and techniques that were used, as well as the concepts, aims and surrounding circumstances. We want to understand why a work came into existence at a particular place and time, and what its significance is. [4]

When the first atomic bombs were dropped on Hiroshima and Nagasaki in August 1945, it was a climactic event that had a profound effect on Albert Einstein's life. In the same instant, the atomic age began and the "innocence of physics" ended. "The first atomic bomb destroyed more than the city of Hiroshima," Einstein commented. "It also exploded our inherited, outdated political ideas." [5]

Sixty years after the bomb was dropped, we do not wish to focus only on his memory. In this book, we wish to focus on some of Albert Einstein's greatest goals: comprehensive military disarmament, the abolition of atomic weapons and the abolition of war. To this day, the Russell–Einstein Manifesto remains a leading document in this context.

Albert Einstein's political life was strongly molded by the East–West conflict. His strategies for peace attempted to shape this conflict after a peaceful fashion, without overcoming it. The resolution of the conflicts during the post-Cold War period, such as asymmetrical wars, conflicts over natural resources, state terrorism, climate change, and poverty, remain the task of science, politics and non-governmental organizations in the 21st century.

The 50th anniversary of the publication of the Russell–Einstein Manifesto in July 2005 obliges us to demand nuclear disarmament in particular and even complete abolition of atomic weapons. This would be a significant contribution to prohibiting the further proliferation of atomic weapons. It also obliges us to work towards the abolition of war.

The Nobel Peace Prize Winner and sole remaining signer of the Russell–Einstein Manifesto, the physicist Joseph Rotblat phrased it vividly: "Either we get rid of the atomic bomb or mankind is in danger of destruction." [6] The largely ineffectual attempts to achieve nuclear disarmament, the senselessness of years of fruitless negotiations, the United States' policy to block arms control entirely on an international level must not lead to resignation and apathy, either in states and governments willing to disarm or amongst parliamentarians and social organizations such as churches and labor unions.

The involvement of the general population and the work of NGOs is indispensable to secure peace. While often still small, weak, isolated, and hardly internationally connected during Albert Einstein's day, these days they are frequently termed the "Second Superpower". Even if the term superpower is a product of modern-day media, the grassroots initiatives of today are a power that influences politics, helps determine public opinion, and can achieve changes in national and international politics. The peace movement has become a power that spans the globe and moves millions of people, often reflecting the will of the majority of the population. Impressive evidence of this power was demonstrated on a new dimension through the resistance to the illegal US-led war on Iraq, when protests reached their peak worldwide on 15 February 2003.

By working together with politicians, for instance the popular support of Mikhail Gorbachev's "new thinking", civil society has helped achieve real atomic disarmament and fundamental societal changes, such as the end of the East–West conflict. Civil society is a seismograph of the recognition of global problems and fundamental shifts. In this sense, Albert Einstein's statement is as relevant today as it was in 1930: "The peoples of the world must take the initiative to prevent themselves from again being led to the butcher block. It is foolish to expect protection from the governments." [2]

Driven away by fascism in Germany, Albert Einstein never again returned to the country of his birth and his many years of work from 1914 to 1933. How profoundly Albert Einstein thought about Germany and the Germans is perhaps shown by his obituary for the heroes of the "Warsaw Ghetto":

> The Germans and the entire German nation are responsible for this mass murder, and as a people they must be punished for it if there is justice in the world. When they are completely defeated and, as after the last war, moan about their fate, one should not deceive oneself a second time, but instead remember that they utterly exploited the humanity of others to prepare their final and most severe crime against humanity. [1]

Early on and unmistakably he warned again and again of the rise of National Socialism. In doing so he came up against a lack of understanding and opposition, especially amongst the German scientific community, including leading representatives and their scientific societies. The seemingly value-free, conservative national German scientists facilitated, consciously or unconsciously, German fascism's rise to power. Albert Einstein's Nobel Prize colleague and co-signer of the Russell–Einstein Manifesto, Linus Pauling, put it this way in 1983, 50 years after fascism took power: "We are not only responsible for our actions, but also for that which we accept without objection." Since fascism in its various historical forms is not only a historical phenomenon, this statement has lost nothing in its relevance for today.

Despite a multitude of prominent requests, e.g. by the Nobel Prize Winner Professor Otto Hahn, Albert Einstein never again became a member of a German scientific society, either of the German Physical Society, for which he served as president from 1916 to 1918, or of the Kaiser–Wilhelm Society successor, the Max–Planck Society. Einstein responded to Hahn:

> It pains me that I must say "no" to you, one of the few men who remained decent and did what they could during those evil years; but I cannot do otherwise. The crime of the Germans is truly the most abominable ever to be recorded in the history of the so-called civilized nations. The conduct of the German intellectuals – seen as a group – was no better than that of the mob. And even now, there is no indication of any regret or any real desire to repair whatever little may be left to restore after the gigantic murders. In view of these circumstances, I feel an irrepressible aversion to participating in anything which represents any aspect of public life in Germany. [2]

In 1954, he granted only one German school permission to bear his name: The Albert Einstein Gymnasium in Neukölln, Berlin. His conscious behavior towards Germany created a personal and a societal rift, especially with "the other Germany" that would have liked to have welcomed him as a comrade-in-arms to a new, antifascist Germany.

Albert Einstein's two letters from 1939 and 1940 addressed to President Roosevelt of the US can only be understood in light of Albert Einstein's profound insight into the criminal nature of German fascism towards the Jews and critics of the regime, and indeed towards all nations of the world who suffered under the German plans to rule the world. In these letters, he warns of the German Reich's possible capacity to create atomic weapons and requests that the US take counter-measures to provide for deterrence against possible German atomic weapons.

From a fundamentally ethically based pacifistic position, there are certainly grounds for criticizing Einstein's actions in 1939 and 1940. Nevertheless, many critics of Einstein's viewpoint underestimate the danger of German fascism's plans to rule the world and how close they were to becoming reality. The dangers to which humanity was exposed by the Nazi regime are confirmed not only by a glance at the map of Europe from 1940, but also by the proximity of German science and research to "attain the bomb." The achievements and capabilities of German science were well known to Einstein. He also knew of Germany's additional material resources attained through its conquests. Letters and documents from his homeland strengthened his fears of "Hitler's bomb". Increasingly, the veil that German physicists had cloaked themselves in after 1945 to play down their deeds is being drawn aside. They were significantly closer to the atomic bomb than they would have had the world believe at the end of the war. Einstein described his own position on the war in this way:

I am today as ardent a pacifist as ever before. Nonetheless, I believe that, until the military threat to the democracies, posed by the existence of aggressive dictatorships, has ceased, we are not justified in advocating the weapon of war resistance in Europe. [2]

Due to the actual historical developments, Einstein regretted his decision to sign the two letters addressed to Roosevelt and called them his worst mistakes. "Had I known that that fear was not justified, I, no more than Szilard, would have participated in opening this Pandora's box." [2]

Einstein worked unflaggingly from 6 August 1945 until his death as a scientist and citizen to fight against any further use of the atomic bomb.

This book adheres to the tradition of Einstein's thinking, namely the undogmatic analysis of the historical and scientific situation and consequent demands on science and politics, as well as the moral stance of every individual, be that person a scientist or simply a citizen. As different and varied as the authors' contributions are, they pose visions and ideals for a more peaceful world: globally, regionally, and locally, for the scientific community, for politics, for society and for the individual.

The purpose of this book is to make a small contribution to the great dream of peace for humanity, in the tradition of Albert Einstein. "But perhaps my voice will contribute towards promoting the greatest of all ideals: good will among mankind and peace on earth." [2]

References

[1] Albert Einstein, *Mein Weltbild*, ed. Carl Seelig, Ullstein, Berlin, 2005.

[2] Nathan Otto, Heinz Norden (Eds.), *Albert Einstein – Über den Frieden*, Peter Lang Verlag, Frankfurt, 1975.

[3] Helen Dukas, Banesh Hoffmann (Eds.) *Albert Einstein, Briefe, Nachlass*, Zürich, 1981.

[4] Peter Galison, Einstein Interview *Ich habe nie verstanden, was Genie bedeutet*, Physik Journal, 3 (2005), 28–29.

[5] Albert Einstein, Thomas Mann, *Letter to the editor, "inter alia"*, New York Times, 10 October 1945.

[6] Ulrich Albrecht, Ulrike Beisiegel, Reiner Braun, Werner Buckel (Eds.) *Das atomare Feuer*, Peter Lang Verlag, Frankfurt, 1986

Opening

Introduction

Mikhail Gorbachev

Last autumn, a large group of scientists and public figures from various countries, among them the author of these lines, came forward with the suggestion to declare the year 2005 "International Einstein Year." Two coinciding anniversaries provided occasion for this celebration. First, 2005 marks 100 years since the publication of the special relativity theory and the quantum theory of light, which have formed the basis for the contemporary scientific world view. Second, it is also the 50th anniversary of the passing of the creator of these theories, one of the greatest scientists of the 20th century – the thinker, humanist and pacifist Albert Einstein.

As this initiative developed, the idea of a book took shape that would be directed to all those who are concerned with the current state of the international community and, in the spirit of Einstein, with a search for a peaceful future for our planet. Two more anniversaries served as motivation for this particular project: 60 years since the destruction of Hitler's fascism in the bloodiest war ever known in history; and 60 years since the dropping of the atomic bombs on Hiroshima and Nagasaki, which in one single moment destroyed hundreds of thousands of people.

This is the book that the reader holds in his hands. Its authors are Nobel Prize winners, distinguished scientists, pioneers of world peace, renowned experts in their field. Here, they consider the problems raised by nuclear weapons, contemplate questions of war and peace, and turn their attention to regional crises and their prevention. We are talking about global approaches, about the survival of humanity, and, in this context, about the environment and the growing shortage of essential resources, such as clean water, etc. Naturally, the authors paid especially close attention to those questions that focus on the cross section of science, technology and warfare, as well

Einstein – Peace Now! Reiner Braun and David Krieger (Eds.)
Copyright © 2005 WILEY-VCH Verlag GmbH & Co. KGaA, Weinheim
ISBN 3-527-40604-2

as the scientific community's involvement in efforts that would ensure peace and disarmament.

Einstein's scientific and moral legacy may be even more relevant today than it has ever been before. His life (1879–1955) coincided with one of the most tragic periods in contemporary history. World War I; the Nazi ascent to power in Germany, Einstein's homeland; the plans for world hegemony devised by the "axis powers," which led up to World War II – these events forged Einstein's convictions as a humanist and a pacifist. At the same time, they also spurred him, at the advice of other scientists concerned with the advances made in atomic warfare by Nazi Germany, to send a letter to President Roosevelt in 1939, urging him to launch the program that became responsible for the development of the atomic bomb in the USA: the Manhattan Project.

As a scientist whose discoveries enabled humanity to master the atom's enormous energy, and along with that, to develop nuclear warfare, Einstein felt the heavy burden of responsibility on his shoulders. During the last years of his life, he showed tremendous concern for peace and overall security. He became one of the founders of the Pugwash movement. Only a week before his death, in an exchange with Bertrand Russell, Einstein agreed to become cosigner of a manifesto that called for new thinking in the atomic age and demanded the rejection of nuclear warfare.

When the famous Russell–Einstein Manifesto was published 1955, it was hard to believe that the ideas of new thinking put forth in the paper would ever be ready for practical use. The Cold War was at its peak. Driven by geopolitical competition, superpowers became ever more entangled in the deadly arms race. Seeing this, Einstein was left with a bitter thought: "The release of atom power has changed everything except our way of thinking."

Some 30 years later, at the outset of the restructuring of the Soviet system that became known as Perestroika, we picked up the idea of new thinking (whose predecessors in Russia had been such scientists as Vladimir Vernadskiy and Dmitriy Sakharov). We wanted to develop this idea and put it in practice, specifically, in order to develop a new approach toward international relations. Several principles crystallized to form the basis of our new political thinking: a) nuclear war is pointless and irrational; b) the concept of war as "an extension of politics, pursued by different means" has lost its relevance and credibility; c) security must be provided largely by political means, including through arms control; d) security is indivisible, it must be provided for everyone, and it must be based on a balance of interests; e) each nation must have a recognized freedom to choose its own path of development and way of life.

One of the first steps on the path toward realizing the principles of new thinking in the realm of international relations became the Soviet–American summit in Geneva in 1985. During this meeting, leaders of both superpowers declared: "Nuclear war is inadmissible; it cannot have a winner." A series of crucial events followed: the Treaty on the Elimination of Intermediate-Range and Shorter-Range Missiles; the Agreement on the Reduction and Limitation of Strategic Offensive Arms; the withdrawal of Soviet troops from Afghanistan; self-determination of the countries of Eastern and Central Europe, which received the freedom to choose their own path; the unification of Germany and the bridging of the European divide. In the end, we achieved the end of the Cold War.

"Peace cannot be kept by force. It can only be achieved by understanding." This is one of the aphorisms ascribed to Einstein. And it is understanding that's lacking in those world leaders who have been infected with the self-righteousness of power and strength. Because of this, the global community regrettably missed many opportunities after the end of the Cold War that should have been used to build a more secure world. Moreover, in addition to the remaining threats, the next period brought new ones.

Today, humanity faces three interconnected global challenges: the challenge of security, the challenge of poverty, and the environmental challenge. International terrorism and local armed conflicts hold the world in a state of continuous strain. Once again, military spending is growing, military technologies are being improved upon, while the weapons trade market is also expanding. At the same time, mass poverty, which has become a fertile breeding ground for violence and extremism, remains a problem on a colossal scale. The environmental situation becomes more and more threatening: global warming is causing changes in our climate, the earth's ice caps are melting, natural disasters are becoming more frequent – and more destructive. Soil is being eroded, thousands of plant and animal species are becoming extinct.

What connects these three challenges are not only their causes, but also their effects, as well as those imperatives, which they dictate to the global community. It will be impossible to resist fanaticism, terrorism and crime, and to ensure global security and stability, without a fight against hunger and poverty. It will be impossible to overcome poverty, without ensuring equal rights to development, fair access to the main means of subsistence, and environmental protection for everyone. Stable and secure development cannot take place without an assurance of peace, even as peace cannot exist without stable development.

Over the last few years, a discussion of development policy at international forums has shown a great degree of agreement on what must be done. Unfortunately, as United Nations General Secretary Kofi Annan rightly points out, we have thus far not reached consensus on how to make our world more secure. Disputes, regarding how best to react to existing threats, reflect deep-seeded contradictions and discord inside the international community.

The contradictions and discord that I am talking about are a matter of fundamental differences. I am convinced that the measures taken toward ensuring international security and fighting poverty will not bring any definitive results, so long as the consumer mentality that exists in the world's most developed nations and is spreading under the influence of globalization, continues to be cultivated. A social model based on consumerism is inadequate at a time when the use of limited natural resources has already exceeded a critical limit.

What's the answer? Does it lie in accumulating military might in hopes of ensuring control over the resources of one's own country and protecting ourselves against the social consequences of the impending environmental catastrophe? That course of action would mean suicide. From the point of view of long-range perspectives, the only possible answer lies in changing our politics. At the end of last year, a high-level panel convened by Kofi Annan and focusing on the threats, challenges and changes facing the world of today, presented a lengthy report, titled, "A More Secure World: Our Common Responsibility." (The panel's existence, coincidentally, means that a suggestion that I have repeatedly put forth to form a kind of alderman's council under the auspices of the UN, has finally found resonance.) The report contains an assessment of the threats that humanity faces today, along with a series of recommendations regarding how best to react to these threats. It also includes recommendations for improving the UN's own structures, including the Security Council.

Nonetheless, many nations, organizations and individuals remain unprepared to respond to the manifest need to change old behavioral patterns. Without bridging that gap, we cannot hope for a radical turn in global politics. Ethical considerations are gaining a key role in international relations and international politics.

The forum of Nobel Peace Prize winners has expressed its opinion on this issue in a statement composed during one of its summits in Rome. According to this statement, politics, if it is to act in the interests of humanity, must rely on ethical values. This will then be reflected in a more just global order, which will be based on the rule of law. The politics of unilateralism and a violent approach toward resolution of conflicts with deep-seeded economic, social and cultural roots, cannot ensure success. Democracy cannot be brought in externally, and certainly not by force. Military doctrine that allows for preventative war and the use of nuclear weapons is unacceptable. These weapons are immoral and must be destroyed. Sophisticated technologies of destruction that have been invented by humanity, must give way to technologies of dialogue, cooperation and solidarity.

In light of this, a special role and responsibility rests with the global scientific community. In paying tribute to Einstein, it is important to emphasize once again the importance of his moral stance as an opponent of Nazism, fascism, all extreme forms of nationalism and militarism, and the politics of violence more broadly. He must be remembered as a consistent advocate of peace, disarmament, human rights, and social justice. To turn to Einstein's tradition of moral thought and to honor his memory may help us to take more effective measures toward rebuilding world politics on the basis of ethics that have their roots in values shared by all of humanity. This volume challenges us to do so.

Mikhail S. Gorbachev

Michael S. Gorbachev was born in 1931. From 1985 until 1991 he was General Secretary of the KPDSU. From March 1990 until December 1991 he was President of the Soviet Union. In 1985 he started the inner Soviet Union reform program, which has internationally become known as "Perestroika" and "Glasnost". For his achievements in easing the East–West conflict and promoting disarmament he was awarded the Nobel Peace Prize in 1990.

How Einstein Became a Politician
Jürgen Renn

Einstein symbolizes like no other the revolutionary character of the science of the 20th century. To the same extent as the sciences themselves, he shaped the public impression of science. This includes the myth of the scientist whose actions are not subject to conventions and authorities. Einstein always used his fame consciously for political issues, but he also used it for the public dissemination of scientific knowledge.

At the start of the 21st century however, the conflicts that gave rise to his nonconformity have fallen into near oblivion. Yet it is these conflicts that explain why the name Einstein is connected with both a significant scientific legacy and with a continued political challenge as well: the necessity for scientists to assume social and political responsibility.

How was it that Einstein became so acutely aware of this responsibility? His entrance into public life is closely connected with his appointment as professor in Berlin. It was during this time that he became a respected intellectual far beyond the realm of science and became a champion of democratic and pacifistic ideals. The contrast between Einstein's Berlin years and the previous period in Switzerland seems astounding at first and virtually inexplicable. For in contrast to Einstein's political involvement in Berlin, no political engagement is found previously, nor did he come to Berlin as a politically active scientist.

At the time of the German invasion of Belgium, Einstein had just been in Berlin for four months. He did not share the general enthusiasm for war and felt isolated in his circle of colleagues due to his attitude towards the daily events. It was not until the infamous "Appeal to the Cultural World" in October 1914 that his political activism was triggered. The Appeal was signed by numerous intellectuals who availed themselves of the ostensible need to protect the German

Einstein – Peace Now! Reiner Braun and David Krieger (Eds.)
Copyright © 2005 WILEY-VCH Verlag GmbH & Co. KGaA, Weinheim
ISBN 3-527-40604-2

culture in order to justify the war and militarism. Einstein did not hesitate to sign a counter-appeal supported by only a few. The consequence of his signature was his further isolation from the academic establishment, but it also brought him into contact with opponents of the war.

How can one individual ever judge societal conflicts and react to them when complex societal mechanisms underlie these conflicts whose structures and laws do not necessarily reveal themselves in their superficial appearance? Activism obviously not only has to do with political and moral issues, it also has to do with cognitive ones. World views and political movements assume the role of mediating authorities as they offer the individual a simplified and, as a rule, false picture of these structures, a picture, however, that often first opens the intellectual resources to an individual so that he may develop his own perspective for action.

In Einstein's case, his family background, experiences with a functioning democracy in Switzerland, and last but not least the implicit political character of his scientific view of himself all played an important role in developing his political views. The origin of this manner in which he viewed himself is especially associated with his reading of books on popular science in which the body of thought behind the bourgeois revolution of 1848 holds particular meaning. But his activism was not provoked until the attempt by leading circles of scientists and artists mentioned above to become directly involved in the political debate, an attempt that in some regards pre-empts the later ideological exploitation of cultural and scientific activities practised by National Socialism. These historical conditions were also accompanied by a dramatic change in the societal role of the scientist, a change that posed a challenge to Einstein. Against the backdrop of his personal experiences, Einstein was able to meet this challenge. Perhaps it was for this reason that the fates of individual persons always remained important to him, as his unflagging actions on behalf of persecuted and condemned persons and emigrants during two World Wars and a Cold War demonstrate.

Jürgen Renn

Jürgen Renn, born in 1956, is Director at the Max Planck Institute for the History of Science in Berlin, Germany. As Director of the MPI Jürgen Renn works on the development of knowledge and its role in society. He became known internationally with his studies about the development of relativity theory.

Part 1
Remembering Einstein

"For a while it is true that our time has accumulated more knowledge than any earlier age, that love of truth and insight that lent wings to the spirit of the Renaissance has grown cold, giving way to sober specialization rooted in the material spheres of society rather than in the spiritual."

Albert Einstein

Einstein – Peace Now! Reiner Braun and David Krieger (Eds.)
Copyright © 2005 WILEY-VCH Verlag GmbH & Co. KGaA, Weinheim
ISBN 3-527-40604-2

World Without War:
A Tribute to Einstein's Quest for World Peace
Joseph Rotblat

Einstein's fame and unique status in the world are mainly due to his scientific discoveries. Much less is known about his political activities: his anti-war campaigns and his advocacy of a world government. Yet, next to science, these matters were nearest to him; he devoted to them much time until the very end of his life.

In this paper on a world without war, I have posed two pertinent questions related to Einstein's peace activities: is a world without war desirable? And, is it feasible? The first question is surely rhetorical. After the many millions of lives lost in the two World Wars of the last century, and in the many wars since, a world without war is assuredly most desirable. And it has been made all the more desirable by the events that have occurred since the end of the Second World War; not only is a war-free world desirable, it is now necessary. It is essential, if humankind is to survive.

I am referring to the development of the omnicidal weapons, first demonstrated in Hiroshima and Nagasaki. The destruction of these cities, heralded a new age, the nuclear age, whose chief characteristic is that for the first time in the history of civilization, Man has acquired the technical means to destroy his own species, and to accomplish it, deliberately or inadvertently, in a single action. *In the nuclear age the human species has become an endangered species.*

Actually, this threat did not loom large when work on the feasibility of the atom bomb began in England, soon after the outbreak of the Second World War. We had a pretty good idea about the terrible destructive power of the bomb. We knew about the blast effect, which would destroy buildings over large distances; we knew about the heat wave, which would consume everything over still greater areas; we envisaged the radioactive fall-out, which would keep on killing people long after the military operations had ended. We even thought of the development of the hydrogen bomb, with its destructive power a

Einstein – Peace Now! Reiner Braun and David Krieger (Eds.)
Copyright © 2005 WILEY-VCH Verlag GmbH & Co. KGaA, Weinheim
ISBN 3-527-40604-2

thousand times greater. But in our discussions about the effects of these weapons we did not for one moment contemplate the ultimate catastrophe that their use might bring, namely the extinction of the human species. We did not envisage this because we knew that this would require the detonation of a very large number – perhaps a hundred thousand – of megaton bombs. Even in our most pessimistic scenarios we did not imagine that human society would be so stupid, or so mad, as to accumulate such obscenely huge arsenals for which we could see no purpose whatsoever. But human society was that insane. Within a few decades, arsenals of that magnitude were manufactured, and made ready for use by the two then superpowers, the United States and the Soviet Union. On several occasions, during the Cold War, we came perilously close to their actual use. I remember, in particular, one such occasion, the Cuban Missile Crisis in 1962, when we were a hair's breadth away from total disaster, when the whole future of our civilization hung on the decision of one man. Fortunately, Nikita Krushchev was a sane man, and he withdrew at the last moment. But we may not be so lucky next time. And next time is bound to happen if we continue with the nuclear policies of George W. Bush.

Morality is at the very basis of the nuclear issue. Are we going to base our world on a culture of peace or on a culture of violence? Nuclear weapons are fundamentally immoral: their action is indiscriminate, affecting military as well as civilians, aggressors and innocents alike, killing people alive now and generations as yet unborn. And the consequence of their use might be to bring the human race to an end. All this makes nuclear weapons an unacceptable instrument for maintaining peace in the world. But this is exactly what we have been doing during, and after, the Cold War. We keep nuclear weapons as a deterrent, to prevent war by the threat of retaliation.

For the deterrent to be effective, the threat of retaliation must be real; we must convince the would-be aggressors that nuclear weapons *would* be used against them, otherwise the bluff would soon be called. George W. Bush, Vladimir Putin, or Tony Blair, must show convincingly that they have the kind of personality that would enable them to push the button and unleash an instrument of wholesale destruction. I find it terrifying to think that among the necessary qualifications for leadership is the readiness to commit an act of genocide, because this is what it amounts to in the final analysis. Furthermore, by acquiescing in this policy, not only the leaders but each of us figuratively keeps our finger on the button; each of us is taking part in a gamble, in which the survival of human civilization is at stake. We rest the security of the world on a balance of terror.

In the long run this is bound to erode the ethical basis of civilization. I would not be surprised if evidence were found that the increase of violence observed in the world – from individual mugging to organized crime, to terrorist groups such as al Qaeda – has some connection with the culture of violence under which we have lived during the Cold War years, and still do. I am particularly concerned about the effect on the young generation.

We all crave a world of peace, a world of equity. We all want to nurture in the young generation the much-heralded "culture of peace". But how can we talk about a culture of peace if that peace is predicated on the existence of weapons of mass destruction? How can we persuade the young generation to cast aside the culture of violence, when they know that it is on the threat of extreme violence that we rely for security?

I do not believe that the people of the world would accept a policy that is inherently immoral and likely to end in catastrophe. This was evident in the reaction to the destruction of the two Japanese cities, a reaction of revulsion, shared by the great majority of people in the world, including the United States. From the beginning, nuclear weapons were viewed with abhorrence; their use evoked an almost universal opposition to *any* further use of nuclear weapons. I believe this is still true today.

On the international arena this feeling was expressed in the very first resolution of the General Assembly of the United Nations. The Charter of the United Nations was adopted in June 1945, two months before Hiroshima, and thus no provision is made in the Charter for the nuclear age. But when the General Assembly met for the first time in January 1946, the first resolution adopted unanimously was to seek the elimination of atomic weapons and all other weapons of mass destruction.

However, from the very beginning, there were hawkish elements among the US leadership, who wanted to maintain a nuclear monopoly for the United States. General Leslie Groves was the overall head of the Manhattan Project, which developed the atom bomb during the Second World War. In October 1945, two months after Hiroshima, he outlined his views on US nuclear policy in a blunt statement:

> If we were truly realistic instead of idealistic, as we appear to be (sic), we would not permit any foreign power with which we are not firmly allied, and in which we do not have absolute confidence, to make or possess atomic weapons. If such a country started to make atomic weapons we would destroy its capacity to make them before it has progressed far enough to threaten us.

During the 60 years since that statement, US policy has undergone a number of changes, but the monopolistic doctrine outlined by General Groves has always been at its base, and now, under George W. Bush, it has become the actual US policy.

During the Cold War years the accumulation of the obscenely huge nuclear arsenals was justified under the doctrine known by the acronym MAD, mutual assured destruction; for each side to have enough weapons to destroy the other side even after an attack. With the end of the Cold War, and the collapse of the Soviet Union, this argument was no longer valid. Then was the time for the abolition of nuclear arsenals, to which the nuclear states are committed under the Non-Proliferation Treaty, signed and ratified by the five overt nuclear weapon states, the USA, Russia, UK, France and China. This, however, did not happen. The United States decided that nuclear arsenals, albeit of smaller size, are needed to prevent an attack with other weapons of mass destruction, such as chemical or biological weapons. And the Bush strategy, partly provoked by the terrorist attack of September 11th went further still; it made nuclear weapons the tools with which to keep peace in the world.

In a reversal of previous doctrines, whereby nuclear weapons have been viewed as weapons of last resort, the Bush doctrine spells out a strategy that incorporates nuclear capability into conventional war planning. Nuclear weapons have now become a standard part of military strategy, to be used in a conflict just like any other high explosive. It is a major and dangerous shift in the whole rationale for nuclear weapons.

The implementation of this policy has already begun. The United States is developing a new nuclear warhead of low yield, but with a shape that would give it a very high penetrating power into concrete, a "bunker-busting mini-nuke", as it has been named.

To give the military authorities confidence in the performance of the new weapon it will have to be tested. At present there is a treaty prohibiting the testing of nuclear weapons, the Comprehensive Test Ban Treaty, which the United States has signed but not ratified.

If the USA resumed testing, this would be a signal to other nuclear weapon states to do the same. China is almost certain to resume testing. After the US decision to develop ballistic missile defences, China feels vulnerable, and is likely to attempt to reduce its vulnerability by a modernization and build-up of its nuclear arsenal. Other states with nuclear weapons, such as India or Pakistan, may use the window of opportunity opened by the USA to update their arsenals. The danger of a new nuclear arms race is real.

The situation has become even more dangerous under the National Security Strategy introduced by President Bush: "To forestall or prevent ... hostile acts by our adversaries, the United States will, if necessary, act pre-emptively."

The danger of this policy can hardly be over-emphasized. If the militarily mightiest country declares its readiness to carry out a pre-emptive use of nuclear weapons, others may soon follow.

Taiwan presents a potential scenario for a pre-emptive nuclear strike by the United States. Should the Taiwan authorities decide to declare independence, this would inevitably result in an attempted military invasion by mainland China. The USA, which is committed to the defence of the integrity of Taiwan, may then opt for a pre-emptive strike.

Altogether, the aggressive policy of the United States, under the Bush administration, has created a precarious situation in world affairs, with a greatly increased danger of nuclear weapons being used in combat.

Ten years after Hiroshima, when we began to appreciate the magnitude of the threat arising from the invention of nuclear weapons, a group of scientists, under the leadership of Bertrand Russell and Albert Einstein, tried to warn governments and the public. Towards the end of his life, as the world situation became increasingly menacing, Einstein's thoughts were ever more focused on achieving a world government system. He did not visualize it as a replacement of existing national governments, but rather as a body with a specific aim: to prevent war by providing the means for solving disputes through negotiation. However, this objective required the relinquishing by national governments of some of their sovereignty – a step that was strongly advocated by Einstein. In the early 1950s, after the development of the hydrogen bomb, grave concern about the growing nuclear threat led to a statement being issued, which has become known as the Russell–Einstein Manifesto. Endorsing the Manifesto was one of the last acts of Einstein's life:

> We are speaking on this occasion, not as members of this or that nation, continent, or creed, but as human beings, members of the species Man, whose continued existence is in doubt.

And it went on:

> Here, then, is the problem which we present to you, stark and dreadful, and inescapable: Shall we put an end to the human race, or shall mankind renounce war?

I am now the sole survivor of the eleven signatories to the Russell–Einstein Manifesto, and as such, it is my duty – even a mission – to keep on posing this question to the public. With the end of the Cold War, and the cessation – for all practical purposes – of the ideological struggle that has polarized the world community – the nuclear threat has somewhat abated, but it has not gone away. The nuclear arsenals have been reduced, but enough warheads are still kept on hair-trigger alert to cause many millions of casualties if set off deliberately, or by a false alarm, or by some other accident. The danger will exist as long as nuclear weapons exist. Robert McNamara, the US Secretary of Defense, during the Cuban missile crisis, expressed this in a simple statement: "The indefinite combination of nuclear weapons and human fallibility will lead to a nuclear exchange."

But even if all the arsenals of weapons of mass destruction were eliminated, the security of humankind would not be assured. Nuclear weapons cannot be disinvented. We cannot erase from our memories the knowledge of how to make them. Should, sometime in the future, a serious conflict occur between the great powers of the day, it would not take long before nuclear arsenals were rebuilt, and we would find ourselves back in the Cold War situation.

Moreover, future advances in science may result in the invention of new means of mass destruction, perhaps even more powerful, perhaps more readily available. We already know about advances in biological warfare whereby gene manipulation could change some pathogens into terrifyingly virulent agents. But entirely different mechanisms might be developed. Just as we cannot predict the outcome of scientific research, we cannot predict the destructive potential of its military applications. All we can say is that the danger is real.

The threat of the extinction of the human race hangs over our heads like the Sword of Damocles. We cannot allow the miraculous products of billions of years of evolution to come to an end. We are beholden to our ancestors, to all the previous generations, for bequeathing to us the enormous cultural riches that we enjoy. It is our sacred duty to pass them on to future generations. The continuation of the human species must be ensured. We owe an allegiance to humanity.

Reaching an agreement on the elimination of the known weapons of mass destruction is very important, because it would remove an immediate source of danger, but in the long run it will not suffice. To safeguard the future of humankind we have to eliminate not only the instruments for waging war, but war itself. As long as war is a recognized social institution, as long as conflicts are resolved by resort to military confrontation, the danger is that a war that begins over a local conflict, for example over Kashmir, will escalate into a global war in which weapons of mass destruction are employed. The probability of this happening at any given time may be very small, but the consequences – should it happen – are so enormous that we must do everything in our power to eliminate the risk. In this nuclear age we can no longer tolerate war, any war. With the future of the human species at stake, this becomes a matter of concern to each of us. A war-free world has become a dire necessity, and its achievement must be made our steadfast objective.

This brings me to the second question in the title of this paper: is a war-free world feasible? To most people, the concept of a war-free world is a fanciful idea, a far-fetched, unrealizable vision. Even those who have come to accept the concept of a world without nuclear weapons still reject the notion of a world without national armaments as being unworkable.

Such attitudes are not surprising considering that, from the beginning, civilized society has been governed by the Roman dictum: *Si vis pacem para bellum* – if you want peace prepare for war. We have paid heed to this axiom despite the fact that throughout history preparation for war has brought, not peace but war. With the onset of omnicidal weapons, the dictum seems to have changed to *Si vis pacem para armas* – if you want peace stay armed to the teeth. Accordingly, both super powers accumulated huge nuclear arsenals in order to keep the peace, and this policy continues now with only one superpower.

The diabolical concept that in order to have peace we must prepare for war has been ingrained in us since the start of civilization. So much so that we have begun to believe that waging war is part of our natural make up. We are told that we are biologically programmed for aggression: that war is in our genes.

As a scientist, I reject this thesis. I see no evidence that aggressiveness is genetically built into our behavior. A group of experts, meeting in Seville under the auspices of UNESCO concluded: "It is scientifically incorrect to say that war or any other violent behavior is genetically programmed into our human nature." In the distant past, under the harsh conditions in which primitive Man lived, he often had to kill for survival, in competition for food or for a mate. Later on, when communities were formed, groups of people killed other groups of people for the same reason, and war became part of our culture. But now this is no longer necessary. Thanks largely to the advances in science and technology, there should be no need for people to kill one another for survival. If properly managed and evenly distributed, there would be enough food and other life necessities for everybody, even with the huge increase in world population. The problem, of course, is that other factors, such as greed, come into play, with the result that the resources are not distributed equitably, and thus many people are still starving, many children are still dying from malnutrition. We have still much to do before the potential for removing the basic causes of war becomes a reality.

Nevertheless, we *are* moving towards a war-free world, even if we do not do it consciously. We are learning the lessons of history. In the two World Wars of the 20th century, France and Germany were mortal enemies. Citizens of these countries – and many others – were slaughtered by the millions. But now a war between France and Germany seems inconceivable. The same applies to the other members of the European Union. There are still many disputes between them over a variety of issues, but these are being settled by negotiations, by mutual give-and-take. The members of the European Union have learned to solve their problems by means other than military confrontation.

The same is beginning to take place in other continents. Military regimes are on the decline; more and more countries are becoming democracies. Despite the terrible bloodshed in recent years – the tribal genocide in Rwanda, the ethnic cleansing in Bosnia and Kosovo, the aftermath of the war in Iraq – the number of international wars is decreasing. We are gradually comprehending the futility of war, the utter waste in killing one another, although this does not seem to apply to terrorists, who show complete disregard for the sanctity of human life.

All the same, for the concept of a war-free world to become universally accepted, and consciously adopted by making war illegal, a process of education will be required at all levels: education for peace; education for world citizenship. We have to eradicate the culture in which we were brought up, the teaching that war is an inherent element of human society. We have to change the mind-set that seeks security for one's own nation in terms that spell insecurity to others.

We must replace the old Roman dictum by one essential for survival in the Third Millennium: *Si vis pacem para pacem* – if you want peace prepare for peace. This will require efforts in two directions: one – a new approach to security, in terms of global security; the other – developing and nurturing a new loyalty, loyalty to humankind.

With regard to world security, the main problem will be preventing conventional wars between nations, and the use of military arms by governments in settling internal disputes. This will require some limitation on the sovereignty of nations, and perhaps a modification of the Charter of the United Nations, which is based on the notion of sovereign nation-states.

Surrender of sovereignty is highly objectionable to most people, but some surrender of sovereign rights is going on all the time, brought about by the ever-increasing interdependence of nations in the modern world. Each international treaty we sign, every agreement on tariffs or other economic measures, is a surrender of sovereignty in the general interests of the world community. To this equation we must now add the protection of humankind.

It is a thorny problem but it has to be addressed. One of the main functions of the nation-state is to ensure the security of its citizens against threats from other states, which is taken to mean possessing the ability to wage war. A change will be called for in this respect: sovereignty will need to be separated from, and replaced by, autonomy. In particular, the right of the state to make war will have to be curtailed. This means no national military forces, and the only legal coercive power on the world scale to be vested in some kind of police force responsible to a global authority. Some form of world governance – as advocated by Albert Einstein – seems a necessary outcome of the evolution of the United Nations.

As a way towards this we have to acquire a loyalty to humankind. As members of the human community, each of us has developed loyalties to the groups in which we live. In the course of history we have been gradually extending our loyalty to ever larger groups, from our family, to our neighborhood, to our village, to our city, to our nation. I should emphasize that loyalty to a larger group is an addition to, not a replacement of, loyalties to the smaller groups. At present the largest group is our nation. This is where our loyalty ends now. I submit that the time has come for loyalty to another, still larger group: we have to develop and nurture loyalty to humanity.

The prospects for developing a loyalty to humankind are becoming brighter due to the growing interdependence between nations, an interdependence not only in the realm of economics, but also in social and cultural matters; an interdependence brought about by the advances in science and technology, in particular, the progress in communications technology; the fantastic advances in transportation, communication and information, that have occurred in the 20th century, and which I have witnessed in my own long life.

Of particular importance is the progress in information technology, in its various forms. The Internet technology enables us to chat with people wherever they are. It provides access to an infinite source of information, and the means to contribute our own knowledge or ideas. Information technology has truly begun to convert the world into a global village: we know one another; we do business with one another; we depend on one another; we try to help one another. We are perforce becoming world citizens.

I welcome the fantastic advances in communication and information as a powerful factor against strife and war, because they provide new means for people to get to know one another and develop a sense of belonging to the whole of the world community.

The applications of science and technology, both the negative and the positive, have created the necessity, and the opportunity to foster the concept of world governance as advocated by Einstein. There is the need for a change in education that recognizes our loyalty to humankind; the need to preserve the human species and the continuation of our civilization.

In the course of many thousands of years, the human species has established a great civilization; it has developed a rich and multifarious culture; it has accumulated enormous treasures in arts and literature; and it has created the magnificent edifice of science. It is indeed the supreme irony that the very intellectual achievements of humankind have provided the tools of self-destruction, in a social system ready to contemplate such destruction.

Surely, we must not allow this to happen. As human beings it is our paramount duty to preserve human life, to ensure the continuity of the human race.

A nuclear holocaust does not appear imminent. Having come close to it on several occasions during the Cold War, we are now somewhat more cautious. But war is still a recognized social institution, and every war carries with it the potential of escalation with fatal consequences for our species. In a world armed with weapons of mass destruction, the use of which might bring the whole of civilization to an end, we cannot afford a polarized community, with its inherent threat of military confrontations. In this scientific era, a global equitable community, to which we all belong as world citizens, has become a vital necessity

In this "Einstein Year", when we honour the great discoveries he made at the start of his life as a scientist, we should also remember his efforts at the end of his life, to create a world without war.

Joseph Rotblat

Joseph Rotblat was born in 1909 and is a Physicist. Until 1944 he worked on the Manhattan Project, which he then left. He is the only living signatory of the Russell–Einstein manifesto, and was the founder and long-term President of the Pugwash Conference on Science and World Affairs. In 1995 he was awarded the Nobel Peace Prize.

1905 was his Great Year – Interview with Hans Bethe

Dieter Hoffmann

On the Death of Hans Bethe

Hans Bethe, nuclear physicist, died on March 6, 2005, aged 98. HOW does the sun shine? It is perhaps one of the first questions a curious child asks about the world, and one that has impelled many a curious child to physics. And, until 1938, it was something that no one could satisfactorily explain. Hans Bethe did so, along with much else, in a long and fruitful career. He was the last of a generation of physicists who changed the world: first in the 1920s and 1930s, by coming up with the entirely new theory of quantum mechanics, and then in the 1940s by proving their relevance in the harshest way imaginable–by creating the atomic bomb. Mr Bethe was famous for his ability to make calculations quickly, a useful talent in the days before computers. It was this that allowed him, in a mere six months, to figure out a problem that had foxed other physicists, and explain what drove nuclear fusion (the process by which two atomic nuclei join and release energy) in the core of the sun. His colleagues could not understand how, as the temperatures of stars increased, they very rapidly became more luminous, so that a star that was ten times as hot would be thousands of times as bright. Mr Bethe saw that only a chain of six reactions, in which carbon acted as a catalyst for the fusion of two hydrogen atoms, would explain what scientists were observing. He was to win the Nobel prize for his calculations.

Related Items

With the advent of the Second World War, Mr Bethe, along with a crowd of other physicists, turned from studying the sun to creating a new one. The Manhattan Project did not start until 1942, three

Einstein – Peace Now! Reiner Braun and David Krieger (Eds.)
Copyright © 2005 WILEY-VCH Verlag GmbH & Co. KGaA, Weinheim
ISBN 3-527-40604-2

years into the war, when Robert Oppenheimer assembled his extraordinary team in the high desert of New Mexico, north-west of Santa Fe, to devise an atomic bomb. One of his first acts was to appoint Mr Bethe head of the theory division, the obvious man for the job.

Among the geniuses working under Mr Bethe was Edward Teller, then one of his close friends. The two men fell out in later years: first about the hydrogen bomb, on which Teller was a hawk and Mr Bethe sceptical, and then about the politically motivated persecution of Oppenheimer, who was stripped of his security clearance. Teller was among those leading the attack, while Mr Bethe steadfastly defended the man who had recruited him. On these matters, and on later attempts to control the bombs he had in part invented, Mr Bethe stood squarely on the side of the angels. It was his duty, he believed, to get closely involved in politics. He helped to get several arms-control treaties ratified in the Senate, most notably the 1963 test-ban treaty, and was early and ardent in his opposition to missile-defence systems, which he said could never be made to work.

He never regretted working on the atomic bomb. When he watched the first test, he wanted only to be sure that the explosion went smoothly; he was not, he said, a philosopher. Unlike Teller, however, he did not become bewitched by his inventions. Instead, he laboured to make sure that they would never be used again.

Hans Bethe talked with Dieter Hoffmann from the Max Planck Institute for the History of Science (Berlin) on 18 September 2004 in Ithaca, NY.

When did you first meet Einstein?

It was 1933, in Princeton, when I was visiting Wigner. He introduced me to Einstein. But the meeting only lasted for two minutes. There was no real exchange. The encounter left no lasting impression whatsoever.

But you'd already heard of Einstein of course?

The first time I heard about Einstein I must have been about twelve years old. Yes, it was in 1918. I was told that Einstein had completely revolutionized physics, but only about ten people in the world understood what he had done. That was wrong of course, because in 1918 his special theory of relativity was both well known and well understood. It's easy to understand! In fact it's one of the easiest works I've ever read. In my opinion the special theory is clearly Einstein's most significant achievement – quite apart from the fact that it's used everywhere in physics. Do you mind if I speak English?

Not at all.

I speak German without any accent, but English more easily. The Special Theory is of utmost importance to atomic motion, but mainly to nuclear orbits, namely the energy in nuclear physics. And also for the observation of nuclear particles coming out of the nucleus. That, of course, Einstein didn't know in 1905, but in my opinion his three papers of 1905 are by far the most important he ever wrote. I think, the Special Theory is far more important than General Relativity.

Even more important than Einstein's light-quantum hypothesis?

Well, the light-quantum is equally important. But there he had Planck as predecessor, whereas the Special Relativity had no predecessor.

As a student and young scientist in the 1920s did you regard Einstein not only as a scientist but also as a political person? (Reverting to German for the rest of the conversation)

For many years, almost up to 1933, purely as a scientist.

Einstein was of course one of the few scientists who said straight away: 'The Nazis are criminals.'

That was my opinion too. Although I wasn't influenced by Einstein, I was in complete agreement with him that the Nazis – judging by their behavior even before they took over power – were criminals.

In 1933 Einstein declared his resignation from the Prussian Academy of Sciences and the Society of Physics because he no longer agreed with the political conditions in Germany. How did you experience Einstein's conflict with the Academy?

I didn't, because in 1933 I was far too busy with my own concerns. As the Nazis classified half-Jews as Semites, it was clear that my future would be entirely different than planned. Fortunately I had a teacher, Arnold Sommerfeld, who supported his students very selflessly, but then met with difficulties himself. By the spring of 1933 Sommerfeld had found me a job in England, with Bragg.

Einstein's relationship with Germany was admittedly very complex. As a young man he fled from the Prussian drill and German school system and went to Switzerland. Then, in 1914, he came to Berlin where he was attacked politically as a Jew and a pacifist in the 1920s. In 1933 he finally emigrated and never returned to Germany. Your attitude was quite different, as you already returned to Germany in 1946.

1948!

You revived old contacts, not only with people but also with institutions.

Only with people, really. I visited Heisenberg and Weizsäcker, and then Gerlach in Munich and Madelung in Frankfurt.

You are a member of the "Order Pour le Mérite" and of German academies. But Einstein refused, in very categorical words, to become a member of the Bavarian and Berlin academies again. After 1945 he always said: 'Germany is the country of mass murderers.' How do you explain your entirely different attitude, not only to the Third Reich but also to Germany in general?

I have always tried to differentiate between the Third Reich and post-war Germany. I knew many Germans – irrespective of whether they were physicists or non-physicists – who could be both honorable and friends. There were of course many convinced Nazis after 1933 and after 1945 as well! But there were at least as many who stayed clear of it.

What role did Einstein play in American society, especially in the American physicists' community? Did people take notice of him?

As a physicist? No. As a political person? Yes. Hardly anyone took any notice of Einstein's physics during those years. There was a small circle of maybe ten people who were working on it and knew about it.

Could you have imagined being one of Einstein's assistants?

Absolutely not.

Why not?

Because he was working on things that didn't interest me.

Einstein signed the famous letter to Roosevelt which addressed the question of using atomic energy for military purposes.

Einstein didn't write the letter. It was written by Wigner and Szilard. They got Einstein to sign because they thought Roosevelt would take more notice of the issue.

As someone who played an important role in atomic physics and the development of the atomic bomb, how do you see Einstein's attitude to the bomb? In 1945 he had become a critic of atomic armament and had developed various ideas on disarmament and a world government.

Most physicists who came to America from Europe shared Einstein's view that the use of the atomic bomb was not only justified but also the only possible means of ending the war with Japan. In my opinion the use of the bomb saved many human lives. But then, in the late 40s and 50s, atomic armament was overdone and that was the opposite of what Einstein, and we all, wanted. It's a very big difference, whether there are two atomic bombs or 200!

Shortly before he died, one of Einstein's last actions involved the manifesto against atomic armament which he signed at the request of Bertrand Russell. What do you remember about this manifesto?

I gave lectures in the same vein. But of course, Einstein was Einstein! People listened to him. I think I was present at two of his lectures. He put things so simply and so convincingly! That's why Einstein was so very important in the propaganda against increased atomic armament. Einstein was an independent individual, independent from his environment.

In science as well as in his perception of public and political affairs?

Yes. Perhaps just one more little anecdote: Einstein and Bohr were involved in one of their many arguments about the interpretation of the quantum theory. Rabi was following this discussion and, although he was of the opinion that Bohr was right – and Bohr was of course right! – he still felt that Einstein was much more self-assured and so had actually managed to win this debate.

Do you have an answer to the question that Einstein also asked himself: What is gravitation?

I'd say that Einstein's General Theory of Relativity is the answer. I don't think it's useful to ask about a graviton. I don't think uniting gravitation with the quantum theory is important. To me they are two completely separate theories. For a long time Einstein thought he could derive the quantum theory from the general theory of relativity. That's impossible of course. The quantum theory is completely independent. There are indeed many physicists who are trying to generalize the modern field theory – using the General Theory of Relativity as the foundation – but I don't think that's the right path.

Nowadays, if you ask people on the street to name a famous physicist, they're guaranteed to say Einstein. How do you explain Einstein's enormous significance and popularity compared with the other great 20th century physicists who were also responsible for major achievements?

I can give you a real reason for that, namely that Einstein changed the fundamental ideas in physics. And that applies to very few of the others. Planck maybe! But Planck never formulated an explicit and logical presentation. And the first person to actually understand the quantum theory was Einstein. Not Planck! And that was in the three publications of 1905. And that's why I'll say it again: 1905 was his great year.

Hans Bethe's Statement Made on the 50th Anniversary of the Hiroshima Bombing:

As the Director of the Theoretical Division of Los Alamos, I participated at the most senior level in the World War II Manhattan Project that produced the first atomic weapons.

Now, at age 88, I am one of the few remaining such senior persons alive. Looking back at the half century since that time, I feel the most intense relief that these weapons have not been used since World War II, mixed with the horror that tens of thousands of such weapons have been built since that time – one hundred times more than any of us at Los Alamos could ever have imagined.

Today we are rightly in an era of disarmament and dismantlement of nuclear weapons. But in some countries nuclear weapons development still continues. Whether and when the various Nations of the World can agree to stop this is uncertain. But individual scientists can still influence this process by withholding their skills.

Accordingly, I call on all scientists in all countries to cease and desist from work creating, developing, improving and manufacturing further nuclear weapons – and, for that matter, other weapons of potential mass destruction such as chemical and biological weapons.

Albert Einstein – A Few Personal Reflections

Walter Kohn

When I was born in 1923, Einstein was 44 years old. His greatest scientific works, three of which we are celebrating this year, were behind him. But much of his impact as a human being on people around the globe was still to come.

Einstein was a simple man, devoid of any sense of self-importance, who, by stripping away inessentials and superficialities, could grasp the core of science, of human nature, of the history of our time, and of the great challenges of the future.

I had the great fortune to come to know him through his friend, Toni Mendel, in the 1940s in Hamilton Ontario. Mrs. Mendel was a well-to-do, highly cultured widow. She had emigrated in the 1930s from Berlin where her home had been a center of lively social and intellectual gatherings which included both Einstein and the great Indian poet, Tagore. (Her son-in-law, Professor Bruno Mendel, is mentioned in the published Einstein–Tagore dialogue.) A deep friendship developed between Einstein and Mrs. Mendel.

During the decade that I and two or three other young people from Toronto visited her from time to time, she was in regular personal correspondence with Einstein, one or two letters per year, each way. The highlights of our visits was her reading from these letters, full of personal warmth and sparkling in their humor and originality.

In 1952 I spent some weeks at the Institute of Advanced Studies in Princeton and Mrs. Mendel kindly provided me with a letter of introduction to Einstein. Our paths occasionally crossed, in the Institute or outside, always with a simple, mutual greeting. But I felt, correctly, that I had not earned the right to take even a few minutes of the great man's invaluable time to formally introduce myself. Later, I was rather ashamed of myself – he would surely have greatly enjoyed the note from his good friend from their common years in Berlin.

Einstein – Peace Now! Reiner Braun and David Krieger (Eds.)
Copyright © 2005 WILEY-VCH Verlag GmbH & Co. KGaA, Weinheim
ISBN 3-527-40604-2

A few years later they both passed away. Alas, Mrs. Mendel's daughter decided to burn Einstein's beautiful letters. But for me and the other young listeners of those times they remained treasured memories, whose spirit has accompanied me ever since.

When in 1979 I became the director of the National Science Foundation's new Institute of Theoretical Physics in Santa Barbara, I was happy that its opening took place on the 100th anniversary of this great man's birth; his thoughts, words and actions have been and continue to be shining lights for us to follow, as scientists or simply as fellow members of the Family of Man.

March 14, 2005

Walter Kohn

Walter Kohn is a Chemist and was born in 1930. He had to leave Austria in 1938. In 1989 he was awarded the Nobel Prize in Chemistry for the development of the density functional theory.

About Albert Einstein

Vitaly L. Ginzburg

Several years ago I had been asked to answer three questions about
Albert Einstein:

1) What do you consider to be the most valuable within the character of scientific activities of Einstein?
2) Which feature in his human and civic image do you like mostly?
   In which episode of his life this feature has manifested most
   brightly?
3) It is very well known that the talent of any scientist is manifested
   most fully if he were born in the particular era for which he was
   predestined. Was Einstein lucky from this point of view?

Instead of answering these questions separately, I have preferred
to say the following. Albert Einstein was an absolutely exceptional
personality, great within great. For me personally even more, he oc-
cupies in general the first position in the history of science and even
in the whole human culture. Here, certainly, it is very important, that
as a physicist I can estimate the principal merit of Einstein – his contribution in physics and, properly, in all natural sciences. Evidently,
for example, a biologist would consider this from the point of view
somebody else, like Charles Darwin.

The creation of general relativistic theory, the decisive role in the
construction of the special relativistic theory, outstanding studies in
the field of quantum theory and statistical physics – all these belong
to Einstein and without this modern physics is inconceivable. When
one speaks about people of such scale, the time of their birth does
not seem to be very important. The great and important problems
were standing and do stand today in front of physics in each epoch.
It can be exceedingly essential for talents, if they were matured and
happened to be at the "right place" in advantageous moments, but
the genius places himself in entirely new ways, although he relies on
his predecessors.

Einstein – Peace Now! Reiner Braun and David Krieger (Eds.)
Copyright © 2005 WILEY-VCH Verlag GmbH & Co. KGaA, Weinheim
ISBN 3-527-40604-2

The unusual repute of Einstein in wide circles, besides his decisive scientific rewards, is also due to his permanently progressive social position and bright publicity talent. Those who know sufficiently well the biography and the epistolary legacy of Einstein, deeply respect and relish his human traits. In 1905, when the famous papers on special relativistic theory, quantum theory and Brownian motion theory by Einstein appeared one after another, he had been earning money for daily living by working for seven long years six days per week, eight hours a day as a technical expert in the patent office in Bern. Is any other similar example known in the history of science?

One can see clearly from the publicity articles and especially from the letters of Einstein, how deeply he understood actual life (and, particularly, the political reality), how widely he knew the history of science and how simple, kind and responsive he was. All that has been said does not mean that Einstein, like other people, never blundered. Yes, he was sometimes wrong in science, and in the assessments of the circumstances of life. But there were very few people, who, in his position, could be so self-critical. For example, at the end of his life, when he was very famous, he wrote to a friend: "You think that I look at the results of my life with the feeling of full satisfaction. In reality everything looks different. There is no one conception, relating to which I would be sure, that it can stay unchanged. I am even not sure that I am on the right path in general".

However, we do know today, and this has been proved by life, by science, that the biggest part of his outstanding life Albert Einstein has been in the right path.

Translated from Russian by Professor Valery S. Petrosyan
(M.V. Lomonosov Moscow State University)

Vitaly L. Ginzburg

Vitaly L. Ginzburg was born in 1916 and is a Physicist. In 2003 he was awarded the Nobel Prize for Physics for his work with superconductors and superfluids.

Part 2

Paths to Nuclear Weapon Free World

"The first atomic bomb destroyed more than the city of Hiroshima.
It also exploded our inherited, outdated political ideas."

Albert Einstein

Einstein – Peace Now! Reiner Braun and David Krieger (Eds.)
Copyright © 2005 WILEY-VCH Verlag GmbH & Co. KGaA, Weinheim
ISBN 3-527-40604-2

Einstein – Man of Peace

David Krieger

Albert Einstein's genius enabled him to peer far into the secrets of the universe, resulting in profound new insights into the relationships between time, space, motion, matter and energy; but he was also passionately involved in the human and social issues of his time as an outspoken critic of militarism and nationalism and a strong opponent of war.

The celebrity that Einstein obtained early in his life through his scientific discoveries gave him a platform from which to advance his social and political ideas. His life encompassed World War I, Hitler's rise to power in Germany, World War II, and the creation and use of nuclear weapons. Throughout his life but especially during the later years he was a fervent advocate for peace and policies that would promote world peace.

When he recognized the terrible destructive potential of the atomic weapons that his theories had, ironically, at least indirectly made possible, he pleaded for a new way of thinking that would make obsolete nation states and their propensity for war and for the establishment of a world government capable of preventing a nuclear cataclysm.

In 1934, Einstein expressed his support for peace in these words:

> The importance of securing international peace was recognized by the really great men of former generations. But the technical advances of our times have turned this ethical postulate into a matter of life and death for civilized mankind today, and made it a moral duty to take an active part in the solution of the problem of peace, a duty which no conscientious man can shirk. [1]

Twenty years later, in 1954, toward the end of his life, Einstein gave this modest assessment of his own efforts for peace:

Einstein – Peace Now! Reiner Braun and David Krieger (Eds.)
Copyright © 2005 WILEY-VCH Verlag GmbH & Co. KGaA, Weinheim
ISBN 3-527-40604-2

In a long life I have devoted my faculties to reach a somewhat deeper insight into the structure of physical reality. Never have I made any systematic effort to ameliorate the lot of men, to fight injustice and suppression, and to improve the traditional forms of human relations. The only thing I did was this: in long intervals I have expressed an opinion on public issues whenever they appeared to me so bad and unfortunate that silence would have made me feel guilty of complicity. [2]

Indeed, time and again Einstein spoke out for what he believed was right. He lived as a human being should live, with freedom and dignity, and sought the same for his fellow humans without regard to borders.

I will review below four important areas of Einstein's thinking on peace: militarism, pacifism, nuclear weapons and world government. I will also examine his thoughts on citizen action.

Militarism

Einstein held the individual human in high regard, but despised the "herd" mentality. "The really valuable thing in the pageant of human life," he wrote, "seems to me not the political state, but the creative, sentient individual, the personality; it alone creates the noble and the sublime, while the herd as such remains dull in thought and dull in feeling."[3] Einstein saw the herd mentality most powerfully reflected in the military system, and he made no effort to hide his disdain for this system. He described the military system as one "which I abhor."[4] He wrote:

That a man can take pleasure in marching in fours to the strains of a band is enough to make me despise him. He has only been given his big brain by mistake; unprotected spinal marrow was all he needed. This plague-spot of civilization ought to be abolished with all possible speed. Heroism on command, senseless violence, and all the loathsome nonsense that goes by the name of patriotism – how passionately I hate them! How vile and despicable seems war to me! I would rather be hacked to pieces than take part in such an abominable business. [5]

Einstein went on to say that he believed this "bogey" of war would have ended long ago had the "sound sense of the peoples not been systematically corrupted by commercial and political interests acting through schools and the Press."[6] All of this was expressed by Einstein in 1931, when he was in his early fifties. Einstein returned to the subject of the herd mentality a few years later, stating, "In two weeks the sheeplike masses of any country can be worked up by the newspapers into such a state of excited fury that men are prepared to put on uniforms and kill and be killed, for the sake of the sordid ends of a few interested parties."[7] He held in particular contempt compulsory military service, which he described as "the most disgraceful symptom of that deficiency in personal dignity from which civilized mankind is suffering today."[8]

Einstein was a strong supporter of human rights and extended the purview of human rights to include "the right, or the duty, of the individual to abstain from cooperating in activities which he considers wrong or pernicious."[9] He believed that this right pertained in the first instance "to the refusal of military service," and cited the Nuremberg Tribunal for the proposition that "conscience supersedes the authority of the law of the state."[10]

In an address to a students' disarmament meeting in about 1930, Einstein told the young people, "The destiny of civilized humanity depends more than ever on the moral forces it is capable of generating."[11] Einstein expressed his belief to the students that the way to achieve a "joyful and happy existence" is through "renunciation and self-limitation," and called upon the youth "to fortify their minds and broaden their outlook through study."[12]

Einstein believed that "the state is made for man, not man for the state."[13] He elaborated that "the state should be our servant and not we its slaves."[14] He continued: "The state transgresses this commandment when it compels us by force to engage in military and war service, the more so since the object and effect of this slavish service is to kill people belonging to other countries or interfere with their freedom of development."[15] Einstein wrote in 1931 that compulsory military service "seriously threatens not merely the survival of our civilization but our very existence."[16]

In an article after World War II, Einstein described the military mentality as making non-human factors essential, "while the human being, his desires and thoughts – in short, the psychological factors – are considered as unimportant and secondary."[17] Among the non-human factors, he listed "atom bombs, strategic bases, weapons of all sorts, the possession of raw materials, etc."[18] Einstein found that in the military mentality an "individual is degraded to a mere instrument," and "naked power" becomes "a goal in itself," a situation he believed was "one of the strangest illusions to which men can succumb."[19]

Einstein's deep opposition to militarism extended to preparations for war. He wrote in a message to a world government meeting in 1946: "You cannot simultaneously prevent and prepare for war. The very prevention of war requires more faith, courage and resolution than are needed to prepare for war. We must all do our share, that we may be equal to the task of peace."[20]

Pacifism

Einstein described himself as "a convinced pacifist."[21] "To my mind," he said, "to kill in war is not a whit better than to commit ordinary murder."[22] He considered Gandhi to be "the greatest political genius" of his time, and believed that Gandhi's efforts for the liberation of India demonstrated "that a will governed by firm conviction is stronger than a seemingly invincible material power."[23]

Einstein viewed pacifism as more than a mere desire for peace or even a refusal to participate in war. "The true pacifist," Einstein wrote to a student group in 1937, "is one who works for international law and order. Neutrality and isolation, when practiced by a great power, merely contribute to international anarchy and thus (indirectly) help to bring about situations that can only lead to war."[24] Einstein expressed a similar opinion in another letter in 1937, this one to the American League against War and Fascism. He wrote: "The supreme goal of pacifists must be the avoidance of war through establishment of an international organization, and not the temporary avoidance of rearmament or involvement in international conflict."[25]

Writing after World War II in 1952, Einstein described the "real ailment" of his time as an attitude created by World War II that "dominates all of our actions."[26] He described this as "the belief that we must in peacetime so organize our whole life and work that in the event of war we would be sure of victory."[27] He believed that this attitude, if not rectified, will "lead to war and far-reaching destruction."[28] Einstein argued that only by overcoming "this obsession" was it possible to reach "the real political problem," which he described as "How can we contribute to make the life of man on this diminishing earth more secure and more tolerable?"[29]

The rise of Hitler, however, caused Einstein to modify his stance. "Up to a few years ago," he wrote in 1934, "the refusal to bear arms by courageous and self-sacrificing persons was such a measure [of pacifism]; it is no longer – especially in Europe – a means to be recommended."[30] It was the fact that non-democratic countries in Europe, such as Germany under Hitler, were basing future plans on military aggression that convinced Einstein that traditional pacifism was no longer viable. "The confirmed pacifist," he wrote, "must therefore at present seek a plan of action different from that of former, more peaceful times."[31]

In Einstein's mind, the advent of nuclear weapons raised terrible new problems for humanity that required bold new and unprecedented action by individual states to prevent war. To achieve this goal, he called for the creation of "a supranational organization supported by military power that is exclusively under its control."[32] Thus, as much as Einstein supported pacifism in principle and opposed militarism, he had arrived at the conviction that military power could not be entirely eliminated. He argued, however, that the control of militarism required supranational control of military power, believing that "[r]eal security is tied to the denationalization of military power," and that this could come about by "converting national armies systematically into a supranational military force."[33]

Nuclear Weapons

Einstein did not consider himself to have had a direct role in the creation of atomic weapons. He wrote:

"I do not consider myself the father of the release of atomic energy. My part in it was quite indirect. I did not, in fact, foresee that it would be released in my time. I believed only that it was theoretically possible. It became practical through the accidental discovery of chain reaction, and this was not something I could have predicted."[34]

One of the scientists who did see the possibility of an atomic weapon, however, was Einstein's friend, the Hungarian émigré physicist Leo Szilard. In the late 1930s, Szilard worried that the Nazis would succeed in developing atomic weapons and this would give them a fateful advantage over the Allied powers. Szilard convinced Einstein to send a letter to President Roosevelt warning of this danger, and it was Einstein's letter that set the US government on the path of developing atomic weapons.

Einstein was not involved further in advising the government on issues related to the bomb. When he heard about the first bomb being used on Hiroshima, he was deeply disappointed and aggrieved. He is reported to have said later, "I could burn my fingers that I wrote that first letter to Roosevelt."[35]

Writing in 1946, Einstein described the changed circumstances brought about by the creation of atomic weapons. "Today the atomic bomb has altered profoundly the nature of the world as we know it," he said, "and the human race consequently finds itself in a new habitat to which it must adapt its thinking. Modern war, the bomb, and other discoveries present us with revolutional circumstances. Never before was it possible for one nation to make war on another without sending armies across borders. Now with rockets and atomic bombs no center of population on the earth's surface is secure from surprise destruction in a single attack."[36] He continued, "Rifle bullets kill men, but atomic bombs kill cities. A tank is a defense against a bullet, but there is no defense in science against a weapon which can destroy civilization."[37]

Alarmed at the frightening prospects of nuclear war, Einstein joined a group of atomic scientists that formed the Emergency Committee of Atomic Scientists in 1946. At a conference held by the committee in Princeton on November 17, 1946, Einstein stated: "Our first task should be to try to communicate to others our conviction that war must be abolished at all costs, and that all other consideration must be of secondary importance."[38] The following statement by Einstein and his fellow trustees of the Emergency Committee was released at the end of the conference:

These facts are accepted by all scientists:

1. Atomic bombs can not be made cheaply and in large number. They will become more destructive.
2. There is no military defense against the atomic bomb and none is to be expected.
3. Other nations can rediscover our secret processes by themselves.
4. Preparedness against atomic war is futile, and if attempted will ruin the structure of our social order.
5. If war breaks out, atomic bombs will be used and they will surely destroy our civilization.
6. There is no solution to this problem except international control of atomic energy and, ultimately, the elimination of war.

The program of the committee is to see that these truths become known to the public. The democratic determination of this nation's policy on atomic energy must ultimately rest on the understanding of its citizens. [39]

In Einstein's view, the atomic bomb represented a quantitative rather than qualitative change in weaponry. He described this change in this way: "The release of atomic energy has not created a new problem. It has merely made more urgent the necessity of solving an existing one. One could say that it has affected us quantitatively, not qualitatively. As long as there are sovereign nations possessing great power, war is inevitable. That is not an attempt to say when it will come, but only that it is sure to come. That was true before the atomic bomb was made. What has been changed is the destructiveness of war."[40]

Einstein did provide more nuance in pointing out that the atomic bomb gave "a considerable advantage in the means of offense or attack over those of defense."[41] He believed that this could lead "even responsible statesmen" to "find themselves compelled to wage a preventive war."[42]

He did not argue in the early years following World War II against the US manufacturing and stockpiling the bomb because he believed that it was necessary for the US to have bombs "to deter another nation from making an atomic attack when it also has the bomb."[43] He believed that deterrence should be "the only purpose of the stockpile of bombs."[44] For the same reason, he believed that "the United Nations should have the atomic bomb when it is supplied with its own armed forces and weapons."[45] At the same time, Einstein believed in a policy of No First Use. "To keep a stockpile of atomic bombs without promising not to initiate its use," he said, "is exploiting the possession of the bombs for political ends."[46] He strongly opposed such exploitation and use of atomic weapons.

In a 1947 article in *Newsweek*, Einstein equated nuclear weapons to the French Maginot Line. "The secret of the atomic bomb," he wrote, "is to America what the Maginot Line was to France before 1939. It gives us imaginary security; and in this respect it is a great danger."[48]

Einstein's answer to nuclear weapons, as was his answer to war as an institution, was supranational organization or world government. He wrote in 1948: "Mankind can only gain protection against the danger of unimaginable destruction and wanton annihilation if a supranational organization has alone the authority to produce or possess these weapons."[49]

World Government

In a 1936 letter responding to a peace plan sent to him by an individual, Einstein expressed his reservations about national governments being trusted to avoid wars. He wrote, "It follows that only a world authority, backed by adequate military power, offers any hope of avoiding war. Indeed, even so radical a step would not guarantee full security. But real protection of peace is certainly not attainable with anything less."[49] In another letter in 1936, Einstein rejected the idea that fascist governments would be willing to explore peaceful alternatives. "I am not in favor of efforts which attempt simply to keep the so-called peace," he wrote. "The only sensible goal today is the creation of an international system of security, unconditionally subordinated to an international authority."[50]

Einstein believed that there was no defense against atomic weapons in either armaments or science or in going underground. The only defense that he believed existed was "in law and order."[51] He wrote in 1946: "Henceforth every nation's foreign policy must be judged at every point by one consideration: does it lead us toward a world of law and order or does it lead us back toward anarchy and death?"[52]

Upon receiving the One World Award in 1948, Einstein told the audience in Carnegie Hall: "There is only one path to peace and security: the path of supranational organization. One-sided armament on a national basis only heightens the general uncertainty and confusion without being an effective protection."[53]

Einstein believed that the path to supranational government lay through strengthening the United Nations. He called for increasing the authority of the General Assembly by making the Security Council and all other UN bodies subordinate to it; for electing delegates to the General Assembly directly by the people rather than by appointing them by governments; and for keeping the UN in session constantly.[54] He believed that "the UN must act with utmost speed to create the necessary conditions for international security by laying the foundations for a real world government."[55]

While favoring inviting the Russians to join a world government, if they were unwilling to join, Einstein advocated proceeding without them, but always holding open the door to them.[56] He thought that the path to world government could be led by the US, UK and Soviet Union. He proposed that a single representative of each government could take on the task of devising a constitution for this government. The principal power of this world government would be over military matters. The only other power that he thought it would need would be "to interfere in countries where a minority is oppressing a majority, and so creating the kind of instability that leads to war."[57]

Einstein said that he feared the tyranny of a world government, but feared "still more the coming of another war or wars."[58] He believed that world government would come, but that it was possible that it could come in the aftermath of a war in which the victor would establish the government under its control. He predicted that such a world government, achieved by military might and war, could "be maintained permanently only through the permanent militarization of the human race."[59] For Einstein, this was a dreaded outcome, and he hoped that ordinary people would lead the way to world government that would be capable of preventing war.

Einstein's views were succinctly expressed in a 1947 message to a gathering dedicated to world law: "Mankind must give up war in the atomic era. What is at stake is the life or death of humanity. The only military force which can bring security to the world is a supranational police force, based on world law. To this end we must direct our energies."[60]

Citizen Action

Einstein placed his hope for the future on the actions of informed citizens. He believed that if citizens accepted the premise of war, there would be a gradual diminishment of civil liberties and that military secrecy would be "one of the greatest obstacles to cultural betterment."[61] The way out for Einstein was achieving supranational government through strengthening the United Nations, which Einstein thought citizens would do if they understood that it was "the only guarantee for security and peace in this atomic age."[62]

He noted early in the Nuclear Age that the determination to avoid atomic warfare was lacking in American leaders and also in its public. "Unless there is a determination not to use [atomic weapons] that is far stronger than can be noted today among American political and military leaders, and on the part of the public itself," he wrote, "atomic warfare will be hard to avoid."[63]

Einstein pointed out that atomic scientists did not believe that the American people could be aroused "by logic alone" to meet the challenge of the atomic era.[64] He argued: "Unless the cause of peace based on law gathers behind it the force and zeal of a religion, it hardly can hope to succeed. Those to whom the moral teaching of the human race is entrusted surely have a great duty and a great opportunity."[65] He called upon all the institutions that molded opinion – including churches, schools and colleges – to "acquit themselves well of their unique responsibility in this regard."[66] He called upon schools to present history "from the point of view of progress and the growth of human civilization, rather than using it as a means for fostering in the minds of the growing generation the ideals of outward power and military successes."[67]

Einstein believed that to insure peace, it was necessary to bring the vital issues to the attention of young people, and that "[t]he spirit of international solidarity...should be strengthened and national chauvinism combated as a harmful force impeding progress."[68]

Despite all the carnage and bloodshed associated with wars that he had witnessed in his life, Einstein retained great faith in humanity.

In his contribution to a college debaters' handbook in 1948, Einstein expressed this faith: "I believe that mankind is capable of reason and courage and will choose the path of peace," a path that he always believed required far more courage than war.[69]

Einstein's Last Legacy

The last public document that Einstein signed before his death in 1955 was the Russell–Einstein Manifesto. The document was written by British philosopher and Nobel Laureate Lord Bertrand Russell, but contained ideas that Einstein had put forward throughout his life. It is, in my opinion, one of the great documents of the 20th century, and remains highly relevant in the 21st century. Concerned about the enormous power of thermonuclear weapons, the Manifesto maintained that humanity had a choice: "Shall we put an end to the human race; or shall mankind renounce war?" Near its end, the signers asked, "Shall we...choose death because we cannot forget our quarrels?"[70]

The Manifesto appealed to scientists and the general public to subscribe to the following resolution:

In view of the fact that in any future world war nuclear weapons will certainly be employed, and that such weapons threaten the continued existence of mankind, we urge the governments of the world to realize, and to acknowledge publicly, that their purpose cannot be furthered by a world war, and we urge them, consequently to find peaceful means for the settlement of all matters of dispute between them. [71]

In addition to Russell and Einstein, the document had nine other signers. One, Joseph Rotblat, the only signer still living, and the recipient of the Nobel Peace Prize in 1995 for his efforts to engage scientists of East and West in the effort to abolish nuclear weapons and war, wrote in 2005:

"Einstein made us think about everything – nothing is absolute, everything is relative. He was a scientist but a realist and aware of what was going on in the world. He was quite the opposite of what people think about scientists – being absent-minded and immersed in their work and naïve. He was fully aware and trying to do something about it. I admire him not only as a great man of science but also as a great human being.

I think if he were still alive, he would still be working on his theories. But he would be working towards peace."[72]

Einstein was a giant among men, who throughout his life stood courageously for peace. He was a scientist who accepted public responsibility for the dangers that science and technology had created. He was never afraid to speak out, nor did he hesitate to do so. Einstein was a citizen of the world, and throughout his life he advocated a government of the world that would bring an end to the institution of war.

Einstein's message to humanity is as relevant today as it was during his life. His is a life not only to be celebrated but also emulated, not least for his deep and abiding commitment to building a peaceful world.

References

1 Albert Einstein, *Ideas and Opinions*, Bonanza Books, New York, 1955, 106.
2 Ibid., 34–35.
3 Ibid., 10.
4 Ibid., 10.
5 Ibid., 10.
6 Ibid., 11.
7 Ibid., 14–15.
8 Ibid., 15.
9 Ibid., 35.
10 Ibid., 35.
11 Ibid., 94.
12 Ibid., 95.
13 Ibid., 95.
14 Ibid., 96.
15 Ibid., 96.
16 Ibid., 98.
17 Ibid., 133.
18 Ibid., 133.
19 Ibid., 133.
20 Nathan, Otto and Heinz Norden, *Einstein on Peace*, Avenel Books, New York, 1960, 1981 edition, 397.
21 Einstein, *Ideas*, 165.
22 Ibid., 165.
23 Ibid., 166.
24 Nathan and Norden, 274.
25 Nathan and Norden, 276.
26 Einstein, *Ideas*, 167.
27 Ibid., 167.
28 Ibid., 167.
29 Ibid., 167.
30 Einstein, Albert, *Out of My Later Years*, Wings Books, New York, 1956, 209.
31 Einstein, *Later Years*, 210.
32 Ibid., 143.
33 Ibid., 144.
34 Nathan and Norden, 350.
35 This comment by Einstein is widely quoted, but without a clear source. See, for example, Richard V. Duffy, *Science Hero: Albert Einstein*, <http://myhero.com/myhero/hero.asp?hero=einstein>.
36 Pauling, Linus, *No More War!*, Dodd, Mead and Company, New York, 1983, 8.
37 Pauling, 8.
38 Nathan and Norden, 394.
39 Nathan and Norden, 395.
40 Einstein, *Later Years*, 185.
41 Ibid., 142.
42 Ibid., 142.
43 Ibid., 193.
44 Ibid., 193.

45 Ibid., 193.
46 Ibid., 193.
47 Nathan and Norden, 404.
48 Einstein, *Later Years*, 155.
49 Nathan and Norden, 271.
50 Nathan and Norden, 273.
51 Pauling, 8.
52 Pauling, 8–9.
53 Einstein, *Later Years*, 147.
54 Ibid., 158.
55 Ibid., 159.
56 Ibid., 196.
57 Ibid., 186.
58 Ibid., 187.
59 Ibid., 199.
60 Nathan and Norden, 407.
61 Einstein, *Later Years* 157.
62 Ibid., 157.
63 Ibid., 194.
64 Ibid., 199.
65 Ibid., 199.
66 Ibid., 199.
67 Ibid., 208.
68 Ibid., 208.
69 Nathan and Norden 469.
70 *Russell–Einstein Manifesto*, http://www.pugwash.org/about/man-ifesto.htm.
71 *Russell–Einstein Manifesto*, http://www.pugwash.org/about/man-ifesto.htm.
72 Rogers, Simon, *The Atom Bomb, Einstein and Me*, The Guardian, January 20, 2005.

David Krieger

David Krieger is a founder of the Nuclear Age Peace Foundation (www.wagingpeace.org) and has served as President of the Foundation since 1982. He is the deputy chair of the International Network of Engineers and Scientists for Global Responsibility (INES). In his early career he was an Assistant Professor at the University of Hawaii and San Francisco State University. He is the author of many studies of peace in the Nuclear Age including *Nuclear Weapon and the World Court*, and has received many awards for his wotk for peace.

The Future of Nuclear Weapons in Europe

Jack Steinberger

Desirability and Feasibility of Nuclear Disarmament

The present nuclear arsenals, a total of some 20,000 warheads, with destructive power on the average each more than ten times the Hiroshima–Nagasaki bombs, constitute a very real threat to humanity. In 2003, on the anniversary of the Cuban Missile crisis, we were reminded by the then US Secretary of Defence Robert McNamara, how close we came in 1963 to this apocalypse. Today, the media worry about the problem of the possible use of nuclear weapons by terrorists, but one rarely hears a word about the much greater danger posed by the nuclear arsenals of the nuclear powers, in particular the US and Russia. I am much more worried about the twenty thousand nuclear weapons Bush, president of a terrorist state, or Putin, without consulting their legislatures, can start flying, than I am about the weapon or two that a "terrorist" may lay his hand on. *The only reasonable, the only possible solution to the menace posed by nuclear weapons is global nuclear disarmament*, and this would be possible, given the complexity of nuclear weapons, if the nuclear powers would resolutely move in this direction, for the greater security of all, especially the nuclear powers themselves.

In contrast with the other great problems humanity faces today: ecological degradation, population expansion, misery and starvation over much of the globe, the problem of nuclear weapons could be resolved rather easily, once the nuclear powers decide that they wish to do this. Unfortunately, this is not the case now for some of these nuclear powers.

Global nuclear disarmament is possible, but this would require the will of the US to lead the nuclear powers toward this end. With US leadership and determination, Russia would follow, China has stated its

Einstein – Peace Now! Reiner Braun and David Krieger (Eds.)
Copyright © 2005 WILEY-VCH Verlag GmbH & Co. KGaA, Weinheim
ISBN 3-527-40604-2

will to follow, India would follow China, Pakistan would follow India, even Israel would have no other choice.

Proliferation of Nuclear Weapons

In 1968 the Nuclear Non-Proliferation Treaty came into effect, signed by all but a few countries. In it, in Article V!, (as clarified in 2000) the nuclear signatories agree to "....an unequivocal undertaking by the nuclear weapons states to accomplish the total elimination of their arsenals leading to nuclear disarmament to which all States parties are committed.....". In the discussion in 1993 towards the renewal of the NPT, there was much friction between non-nuclear and nuclear states concerning the future of Article V!, since in fact the nuclear arsenals had not been dismantled as agreed to in Article V!, but instead were now considerably larger than in 1968. India declared that its continuation as treaty member would be conditional on a fixed time limit on total nuclear disarmament, but this was refused by the US, so that five years later we had two additional nuclear powers, India and Pakistan. As time progresses, it is more and more difficult for the non-nuclear states, especially those outside the western community, to accept their unequal status under this treaty. This constitutes a strong force, driving proliferation.

Evolution of United States Nuclear Weapons Policy

Historically, the country driving the escalation of the nuclear arsenals has been the US, followed by Russia. Present arsenals are roughly: US 10,000 warheads, (3000 Megatons, 200,000 Hiroshimas), Russia 7000 warheads, England, France, China of the order of 400 each, Israel, India, Pakistan, I don't know, perhaps of the order of dozens or 100 each. During the Cold War the US position was that the augmentation was driven by the Soviet Union, although the facts were clearly the opposite. Since the collapse of the USSR, the US and Russia have reduced their arsenals by about a factor of two, mostly eliminating the tactical weapons, but US nuclear weapons policy maintains that thousands of nuclear weapons are still essential to its defence, whether or not other countries possess nuclear weapons. In order to encourage the maintenance of the NPT, President Clinton promoted and signed the Comprehensive Nuclear Test Ban Treaty, which however remains unratified by the US congress. Instead, under the present administration, the US has embarked on a program to develop a new type of "usable" nuclear weapon, the so-called "bunker buster" that it proposes to test. Resumption of nuclear testing would be an extreme blow to nuclear disarmament; it would signal a resumption of the nuclear arms race, and encourage proliferation. In the 1960s it was realized that anti-nuclear missile defence would only encourage nuclear escalation, since it is cheaper to build more bombs than antimissile missiles, with the result that the US and the USSR signed an antimissile treaty. Two years ago the US withdrew from this treaty, thereby encouraging an increase in nuclear armament, especially by China and Russia. The position of the US with respect to nuclear disarmament has been poor from the beginning; it is extremely disappointing in recent years. The US maintains that its nuclear arsenals are essential to its defense, and asserts the right to "pre-emptive" use of nuclear weapons also against possible non-nuclear activities that it may consider as a threat. It is difficult for me to understand this policy, since, given its overwhelming superiority in conventional weapons, as best I can understand, the US would be much more secure, also against terrorism, if there were no nuclear weapons, and so would have the most to gain by global nuclear disarmament.

What can Europe, Germany in Particular, Do to Promote Global Nuclear Disarmament

Clearly, the leadership for nuclear disarmament must come from the US, but perhaps the rest of the world can nevertheless do something significant, by calling attention to the problem and pointing to its solution. The gesture I have in mind is linked to the related problem confronting Europe, that of making its defence coalition, NATO, more independent of America. From its inception, in the shadow of the "Cold War" confrontation with the USSR, and given its military superiority, the US has dominated NATO, and has seen NATO as its means to dominate European military policies. With the collapse of the Soviet Union, and in the light of the Iraq war, it is clear that it is the interest of Europe, while maintaining friendly relations with the US, to be reasonably independent, not to follow the US "im gleichen Schritt und Tritt".

During the Cold War, US nuclear weapons stationed in Europe, within the frame of NATO, numbered about 10,000, almost all "tactical" (rather than "strategic"). The tactical weapons, in agreement with Russia, have now been largely eliminated, but the principle of "nuclear arms sharing" within NATO is maintained. Nuclear weapons are stationed in Germany, England, Belgium, the Netherlands and Italy, perhaps some 500 B-61 nuclear bombs altogether, with some 150 in Germany [1], the actual numbers are not openly stated. In case of war, they and their possible use would be under the responsibility of the country is which they are stationed. This is in clear violation of the NPT. It is also important to understand the political role of the consequent "shared burden of responsibility" for nuclear forces by the European host. The chief function of these nuclear weapons is symbolic: the US "nuclear umbrella", or military dominance, over Europe, and consequent support for US policies for the maintenance of nuclear arsenals, against global nuclear disarmament, in violation of the Non-Proliferation Treaty.

Do these weapons, does the US nuclear umbrella, increase European security, or the opposite? A clear statement by the European countries in which these weapons are stationed, in particular by the German government, that it believes that the world would be better, more secure, without nuclear weapons, that the time has come for a plan, a program for global nuclear disarmament, that it therefore no longer wishes nuclear weapons stationed on its soil, would be a forceful act; it would remind the world of this grave problem, and give an impetus toward global nuclear disarmament. Some members of the government might fear that such an initiative might harm German–American relations and possibly the German economy, but this is not necessarily the case. Not only does Germany benefit from good relations with the US, but so does the US need the good relationships of other governments. Some dozen years ago Spain took such a step, nevertheless Spanish relations with the US seem as good as Germany's, and the Spanish economy is surviving. While this gesture would not resolve the problem of the nuclear arsenals, it would at least be a significant step in the right direction.

Reference

1 Robert S. Noris and Hans M. Kristensen, *US Nuclear Weapons in Europe*, 1959–2004, in: Bulletin of the Atomic Scientists, Nov./Dez. 2004, pp. 76–77.

Jack Steinberger

Jack Steinberger, born in 1921, is a Physicist. For more than 30 years he has worked as a Scientist for CERN. In 1988 he was awarded the Nobel Prize for Physics.

Einstein, Peace and Nonproliferation: a Latin American Perspective

Ana María Cetto and Luis de la Peña

Einstein the pacifist

We would like to begin our contribution with a topic that for some authors of the present book may be well familiar, but that is much less known or even misrepresented among the broader public, who will hopefully constitute the main readership of this volume.

One frequently finds statements in the popular press as well as among the population, to the effect that Einstein was directly responsible for the production of the atomic bomb, even to the extent of considering him the father of the bomb, not only intellectually, but even materially. This special Einstein Year gives us, among many other things, a renewed opportunity to clarify these matters. This task is important because the high moral and human values that accompanied Einstein throughout his entire life constitute one of the best models that can be offered to our young people. Mystifying, distorting or forgetting the facts undermines the very objectives and values that guided Einstein in his lifelong pacifist efforts. It is also important because in the world of today we continue to witness the recourse to force and menace, even nuclear, as a more than frequent means to deal with international or local conflicts.

There are two common misconceptions about Einstein's participation in the making of the nuclear bomb. One is that his famous law (undoubtedly the most popularly known physical law) that energy equals mass times the squared velocity of light, is the direct cause of the bomb. The other is that he directly participated in its making. These two assertions imply that Einstein carries a huge and direct responsibility for the bomb, its consequent use against Japan, and the arms race that followed with all its implications for human beings and the biosphere. These beliefs are so distant from the bare facts and from the deep motivations and convictions that directed Ein-

Einstein – Peace Now! Reiner Braun and David Krieger (Eds.)
Copyright © 2005 WILEY-VCH Verlag GmbH & Co. KGaA, Weinheim
ISBN 3-527-40604-2

stein's actions throughout his life, that every opportunity to throw light on them should be seized.

Of course it is true that Einstein discovered that mass and energy are only two different forms of the same physical entity, so that they can be transformed into one another, with a consequent delivery of tremendous quantities of energy from small masses. Precisely the last paper produced during his *anno mirabilis* was the one in which he presented the derivation of his now famous formula $E=mc^2$ as a by-product of the theory of relativity just proposed in a previous paper of this astounding series. But that was 1905. No shadow of any possible civil or military use of his discovery could be envisaged. Even Einstein himself did not speak of extracting energy from mass in that 3-page paper. To consider that he worked for the military by publishing this result would be equivalent to saying that Newton, by discovering the law of gravitation, is to be blamed for the ballistic missiles. With his law, Einstein was making a profound scientific discovery, a conceptual breakthrough in the understanding of the physical world.

The novelty in Einstein's law relating mass and energy was absolute. Before his 1905 paper, mass and energy were two totally independent and unrelated entities. There was a law of conservation of mass and a law of conservation of energy during the physical (and chemical) processes, but each law dealt with a different thing. On top of the huge surprise of the unexpected mutual interconvertibility of mass and energy, was the predicted enormous quantity of energy that appeared associated with a given mass. The new knowledge opened doors to very important discoveries, particularly in the realms of nuclear physics, elementary particles and astrophysics. Among these we especially note the solution of a very old and most intriguing question, namely that of the source of energy of our sun (and the other stars, of course).

Even long thereafter, during the late 30's, Einstein as well as many other scientists, including the great Danish physicist Niels Bohr, were still convinced that the bomb, if at all possible, was many years ahead, as tremendous technical difficulties would be inherent in its production. It was not until a succession of discoveries made during the decade of the 30's that it slowly became clear that a nuclear chain reaction delivering enormous energies by the conversion of a small fraction of the mass of the nucleus of uranium into energy was feasible. 'Chain reaction' expresses the notion that the physical conditions required to produce the nuclear reaction are reproduced with some gain in each step, so that every reaction gives rise to more than one successive reaction in a very fast process, releasing the whole of the accessible energy in a small fraction of a second. Once this became clear, several physicists, among them the Hungarian Leo Szilard, first looked for means to stop all research that could lead to the bomb and, after failing in such attempts, to redirect the research in the best possible direction, given the terrible circumstances in which the world was involved.

The name of Leo Szilard has not acquired widespread eminence among the general public, although he deserves all our consideration and esteem, both as scientist and as a highly responsible human being. As was the case with many other scientists and intellectuals, Nazi persecution of the Jewish people forced him into exile. He was the first to conceive the possibility of attaining a nuclear chain reaction. Since by then (1936) Hitler's Germany was in a fast and alarming process of re-arming, he patented the procedure in England and assigned it to the British Admiralty to keep it in secrecy, being afraid of its possible military use. However, sooner rather than later the news about the feasibility of the uranium chain reaction spread. Things became worse as the Germans invaded Czechoslovakia, the country with the second largest reserve of uranium ore in the world, and prohibited exportation of this ore. They would soon invade Belgium, the country with the largest uranium reserve of all. For Szilard this meant that most probably the Nazis were already after the bomb, with all its foreseeable threat to humanity. Here we find the root of the famous Einstein letters to President Roosevelt pressing him to create a Committee for the study of the uranium chain reactions.

In search of effective means to avoid the possibility that an eventual nuclear bomb of devastating effects could be in Nazi hands alone, Szilard looked for Einstein's help. By that time also Einstein had been forced to leave Germany and was living in the United States. They were good old friends, had worked together in Berlin on various projects and even had jointly registered some patents. However, more important was that they held the same views about the tragic meaning for humanity of an eventual victory of the Nazi forces in Europe. Szilard visited Einstein (accompanied by another important Hungarian physicist, Eugene Wigner) to propose to him that he send a letter to President Roosevelt. This letter would explain the real possibility of constructing an uranium based bomb, the apparent efforts of the Germans to reach it, and the almost certainty that this could be achieved in the immediate future, that is, in time to be used during the war. In line with his pacifist principles, Einstein accepted and signed the letter drafted by Szilard; the letter was sent in October 1939. [1]

Roosevelt did indeed appoint a Committee to study the problem. However, its work was too slow, probably because the production of a bomb seemed too remote to be ready before the end of the war. Einstein and Szilard (joined by Alexander Sachs, a friend of them and of Roosevelt) sent a second letter in April 1940 to the President. Things did not change much until more than a year afterwards, when a British official report was presented to Roosevelt, stating that the bomb was feasible and could be reached in a couple of years. [2] Then things changed abruptly; Roosevelt approved the Manhattan Project and, by December 2, 1942, Enrico Fermi and Leo Szilard achieved the first controlled nuclear chain reaction in Chicago. All that was needed to produce the bomb was to lose control of the reaction by increasing the purity of uranium 235 (the fissionable isotope) from some few percent to above 90 %. Simple as it sounds, to achieve this for a few kilograms of uranium (the critical mass required to sustain the reaction) was a formidable technical, economical and managerial problem. For the first US bomb this task alone demanded hundreds of thousands of workers laboring for several months. Almost as many people participated in the making of the bomb as the number of its victims – a human tragedy of huge dimension, both physical and moral.

Einstein made no further interventions in the making of the bomb. Even more, it was decided to keep him distant from the project, since its leaders did not trust that he would maintain the required secrecy. It should be emphatically stressed that Einstein was forced by the severity of the war situation to change his deep and life-long pacifist and antimilitarist views, his proactiveness for disarmament and against any form of war and the narrow nationalism that fosters it. He called all his life for peaceful solutions to any kind of problem. His declaration "I would unconditionally refuse all war service, direct or indirect, and would seek to persuade my friends to take up the same stance, regardless of how I felt about the causes of

any particular war", represents him well, and thus though never published, found its way into the international press. [3] Confronted with
the Nazi atrocities he was forced to change. Already in the early 30's
he declared: "I loathe all armies and any kind of violence; yet I'm
firmly convinced that at present these hateful weapons offer the only effective protection". Should Nazi militarism prevail, "you can be
sure that the last remnants of personal freedom in Europe will be destroyed" (Ref. 3). With the maximum possible regret, he was forced
by the Nazi blatant menace to humankind to promote once in his life
the only possible deterrent at that moment. After the war he declared: "If I had known that the Germans would not succeed in constructing the bomb, I would never have lifted a finger". Even more,
once the German menace was over there was a third common letter
to President Roosevelt (as the previous ones, written by Szilard) asking him not to use the bomb on moral grounds. This letter remained
unopened in Roosevelt's desk after his sudden death.

This short and schematic account shows that Einstein's letters
were of little effect in the sense that, even without them, the United
States would have set out to produce the nuclear bomb. But one may
speculate that they contributed to accelerate the pace, thus making it
ready for use in the war against Japan. [4] Surely this thought did not
escape Einstein and in fact embittered the latter years of this convinced pacifist. Indeed, he declared a few months before his death
that signing the letter to President Roosevelt was the great mistake
of his life, "but there was some justification – the danger that the
Germans would make the bomb."

After the war and until his death Einstein dedicated most of his efforts and time to campaign for peace and disarmament, with a strong
emphasis on the dangers of the arms race in a nuclear world. He in-
sisted once and again, in different ways: "War is a terrible thing, and
must be abolished at all costs"; "war cannot be humanized. It can only be abolished."

Szilard's attitude to the bomb was much the same as Einstein's for
the rest of his life (Einstein died in 1955 and Szilard in 1964). Andrei
Sakharov is a third example: he was central for the development of
the Soviet hydrogen bomb, but afterwards and until the end of his
life became a strong opponent of nuclear weapons and the arms race.
After the war, Szilard even switched to biophysics to veer away from
nuclear physics. From 1944 on he worked openly against the use of

the bomb, even warning President Truman about the dangers of its use against Japan. He organized a successful movement to keep atomic energy separate from military control, and opposed the development of the hydrogen bomb and the nuclear testing. He founded, with Einstein, the Emergency Committee of Atomic Scientists, which started publication of the important anti-nuclear Bulletin of the Atomic Scientist, and participated from the beginning in the Pugwash Conferences (initially summoned by Einstein and Bertrand Russell) that opened important lines of communication by gathering scientists from the East and the West to look for means to reduce tensions and promote the nuclear disarmament.

Immense nuclear arsenals still exist today (more than 30 000 warheads) and, worse than ever, the knowledge to build the bomb is extending and continues to develop, while fissile material is becoming reachable through an underground market. Thus the legacy that Einstein and many others gave to us is still alive and calls all of us, around the world, to take open action to further peaceful solutions to our problems and conflicts. No arms are needed in a peaceful world.

These considerations open the door to our second comment, on the sustained efforts deployed by countries in Latin America to banish nuclear weapons from the region and from the face of the earth.

Latin America and nonproliferation

In 1967, at the height of the Cold War, the Treaty of Tlatelolco was signed by states in Latin America, creating the world's first regional Nuclear Weapons Free Zone (NWFZ). This Treaty commits the parties to use nuclear power for peaceful means; they are required to prohibit and prevent the testing, use, manufacture, production, acquisition, receipt, storage, installation, deployment and possession of nuclear weapons in their territory. To ensure its effectiveness, the Treaty includes two additional protocols committing states with responsibility for territories in the region (France, Holland, the UK and US), and the major nuclear powers (China, France, Russia, UK, and US) to maintaining the Zone. [5]

The creation of an extensive NWFZ is no small feat, and speaks for the normative power of the Treaty. Over the past 36 years we have seen the transformation of a region from one with several grave emerging nuclear threats to a truly Nuclear Weapons Free Zone. One should have in mind that the Treaty was signed only five years after the Cuban Missile Crisis, at which time Cuba remained committed to maintaining the option of nuclear weapons as long as its conflict with the US persisted. In addition, Argentina and Brazil were engrossed in their own race for nuclear arms during the 1970s and 80s. Finally, following Cuba's ratification in 2002, all 33 states in the region have signed and ratified the Treaty.

The agency responsible for ensuring compliance with the Tlatelolco Treaty, OPANAL, has been actively engaged in promoting nuclear non-proliferation both in Latin America and around the world. Its XVIII General Conference that took place in Havana in 2003, the first one since the Treaty entered into force for all states, addressed important outstanding topics, such as the transport of radioactive materials and the related problem of radioactive contamination of the marine environment, the role of Nuclear Weapons States (NWS), and the dilemma posed by possible US transport of nuclear weapons in the hemisphere.

The role of the NWS in the Treaty regime is one delicate issue. China has stated that it will not be the first to use nuclear weapons and will not threaten or use nuclear weapons against Latin America. France, in its turn, asserts that its ratification of the protocol is understood not to be an obstacle to the right of self defense; this statement leaves room for the use of nuclear weapons in self defense, even against non-NWS. Finally, the declarations of Russia, the UK, and US are related to self-defense but have even broader applications. All three countries reserve the right to use nuclear weapons against Latin American states in the event that those states commit acts of aggression with the support of a NWS. In 2003, the Secretariat of OPANAL sent notes to all five countries requesting that they review and consider withdrawing or modifying their declarations. During the XVIII General Conference, China, France, the UK and Russia indicated that they are studying the request, and reiterated their continued support of the Latin American NWFZ. The US made no statement.

The State Parties to the Treaty have also been working since 1996 to create an international conference of all Nuclear Weapons Free Zones around the world. This is an ambitious project; yet in 1967 the creation of a NWFZ in Latin America was itself highly ambitious, and much has been accomplished since then.[1]

Let us look at the recent history of the Latin American States with regard to the Nuclear Non-Proliferation Treaty (NPT). There are some important countries who have joined the NPT since 1995, specifically Argentina and Brazil, who were and could be in principle nuclear weapons capable. In November 2002 also Cuba submitted its instrument of ratification to the NPT, as a demonstration of its political will to support international arms control agreements. Cuba's accession leaves India, Israel, and Pakistan as the only countries that are not party to the Treaty.

Take in particular the case of Brazil: the country had initiated a national nuclear programme as early as 1940, allowing the US to mine its uranium reserves (the world's sixth-largest ones) in return for nuclear technology. Still in 1978, Brazil signed an agreement with West Germany to obtain 'significant' nuclear technology. Yet in 1980, Brazil and Argentina, which had also developed a nuclear capacity, decided to establish a cooperative agreement on the peaceful use of nuclear energy (ratified in 1994), which was followed in 1985 by a reciprocal safeguards agreement. In 1990 the Brazilian government symbolically closed a nuclear test site and exposed its atomic weapons plan. Recently Brazil granted IAEA inspectors limited access to crucial uranium-enrichment technology, to ensure that it is only enriching uranium to the low levels needed for civilian nuclear reactors rather than to the higher levels that could provide the explosive material for a nuclear weapon. (Such limited monitoring was justified on the grounds that opening its facility to inspections could lead to industrial espionage of what is claimed as novel enrichment technology.) This step comes as Brazil is taking on an increasingly high international profile. Amb. Sergio Duarte will chair the 2005 NPT Review Conference, bringing together all 189 States Parties to

1) In April 2005, on the eve of the NPT Review Conference, the first Conference of the NWFZ finally took place in Mexico City, with participation of 91 State Parties of the four NWFZ established so far plus 38 observer countries. The Conference made a strong appeal to the NWS to act on their 2000 NPT commitment to an "un-equivocal undertaking" for a nuclear-weapon-free world. As is well known,, regrettably, the NPT Review Conference itself failed to make any progress to strengthen the Non-Proliferation Treaty and was even unable to adopt a consensus final document on substance.

review the last five years of the Treaty and make recommendations for its continued implementation. Brazil also is a leading member of the New Agenda Coalition, the grouping of eight non-nuclear nations launched in 1998 with the goal of pressuring the NWS to fulfill their NPT disarmament obligations.

Would it be surprising that we Latin Americans generally maintain a critical perspective of nonproliferation? In our region there were countries perfectly able to become nuclear weapon states but have chosen not to, giving up unilaterally their nuclear weapons capability, for the benefit of the whole. Let us remember also that in the case of the African NWFZ there was a country that gave up not only its nuclear weapons possibilities but its nuclear weapons status. South Africa dismantled its incipient nuclear arsenal, once again to the benefit of international peace and security. The NWFZ represents an international moral and diplomatic example to be followed, when there is real political will. More than 110 countries have already chosen this political path, and the non-NWS Parties to the NWFZs have proven that real nuclear disarmament is possible.

But, as has been forcefully stated by Amb. Román-Morey, [6] the Nuclear Weapons States should not take the NWFZ for granted. Nuclear disarmament is a joint venture, it is not only the "Disarmament of the Unarmed" as once Latin America was taken to be. The disarmament efforts should be carried out in full co-responsibility between nuclear and non-nuclear weapons states. We just have to keep in mind that the tragic results of a nuclear conflagration would affect our world in its entirety, because nuclear weapons have no destruction limits, no geographical lines, no political borders, no legal reasons.

When the NPT entered into force in 1970, sensitive nuclear technology was widely considered out of the reach of most countries. This is no longer the case. Access to such technologies has increased, particularly over the last few years. As many as 40 countries may now have the technical know-how required to produce nuclear weapons, and the legal and political regimes have not kept pace with these technological developments.

Article IV of the NPT has two interconnected elements. The first reaffirms the right of all NPT parties "to develop research, production, and use of nuclear energy for peaceful purposes, without discrimination." The second is a reaffirmation that "all the Parties to the

Treaty undertake to facilitate and have the right to participate in the fullest possible exchange of equipment, materials and scientific and technological information for the peaceful uses of nuclear energy" and places an obligation on the parties to "cooperate in the development of nuclear energy for peaceful purposes", especially in the territories of non-NWS Parties, with due consideration for the needs of the developing areas of the world.

There is the widespread view that the NWS have backed away from their original guarantee that the non-NWS would enjoy "the right to participate in the fullest possible exchange ... for the peaceful uses of nuclear energy", as well as the right to receive assistance from "Parties to the Treaty in a position to do so". The increasing concerns that those "in a position to do so" are not only no longer doing so but are placing still more restrictions on supply, have fostered a belief among many non-NWS that the NPT bargain is being corroded. [7]

The situation is now compounded as the illegal nuclear market continues to develop. As stated recently by M. ElBaradei, the IAEA's Director General, "some countries make schemes and others produce centrifuges that further go through a third party to undisclosed ultimate consumers. Nuclear experts, dishonest companies and even governmental agencies are involved here." [8]

The new challenges to the nuclear nonproliferation regime cannot be met effectively with the old Cold-War mentality, they require a fresh response. In particular, a new look needs to be taken at multilateral approaches to the various phases of the nuclear fuel cycle. Such new approaches are clearly in their infancy, and their progress depends mainly on the political will, first and foremost of the nuclear powers.

This brings us to the question of how far one can go in furthering such multilateral approaches without a genuine reform of the multilateral system as a whole. Can there be an effective nonproliferation process if we do not have an adequate multilateral system for our time? Does the political willingness exist that is needed for a thorough and honest revision of the workings and mechanisms of the current system?

The gap between the so-called First World and Third World, between North and South, is a flagrant demonstration of the serious inadequacies of the present world system or 'order', as some like to call it. The fact that this gap continues to show signs of widening, is a demonstration of the urgent need for such a thorough revision. Not only nuclear non-proliferation is at stake, but more broadly, the survival and wellbeing of a large part of the population on earth. Take, in particular, the Latin American hemisphere, where 44 % of the population are marginalized and one-tenth of the population must live on 1.6 % of the total income. Under this stress, peaceful, nonviolent solutions to conflicts are difficult to sustain for long.

Latin America's proactive policy against nuclear weapons has been a source of pride for us. It represents the outcome of a significant collective effort that has brought together the wills of heads of government and decision makers in the region. It has been supported and promoted by a diversity of nongovernmental organizations, involving intellectuals, scientists and activists in different fields. But beyond that, it is grounded on a widespread tradition of peace and nonviolence of our peoples.

This culture of peace against all odds has been furthered by nonviolent, antimilitarist movements involving among others peasants, workers, priests, professionals and students, who work with the marginalized people to bring them into full participation in the economic, political and social life in their countries. Latin America has had its share of revolutions and aggressions. However, given the high levels and long duration of violence experienced by its peoples, the extent of commitment to nonviolence has been impressive. [9] Unfortunately, recent signs of extreme urban violence in several countries may be a dangerous indication that people's patience is reaching its limits.

Five trillion dollars have been spent since the start of the nuclear age on the development, manufacture and deployment of nuclear weapons. [10] These weapons continue to be a menace for the world as a whole, despite considerable political and diplomatic efforts from many countries. But our world today is not only a nuclear world, it is one in which globalization develops and extends to all human activities. This means, amongst many other things, that all kinds of interrelations among economies are being deepened and that otherwise local problems become global, magnifying the dangers caused by conflicts and wars. Thus the mechanisms of defense and preservation of peace should be simultaneously strengthened and generalized so that more effective and powerful means can be put into motion with due swiftness. In particular, the United Nations rules and mechanisms formulated fifty years ago for an entirely different world require an urgent and profound revision if this multilateral system is to preserve a capability of international effectiveness when confronted with exacting problems.

References

[1] Spencer Weart and Gertrud Weiss Szilard, (Eds.), *Leo Szilard: His Version of the Facts; Selected Recollections and Correspondence* (Cambridge, MIT Press, 1978).

[2] Richard Rhodes, *The Making of the Atomic Bomb*, Simon and Schuster, New York, 1988.

[3] A brief and highly recommendable popular account of the pacifist and antimilitarist views and struggles of Einstein is given in: www.ppu.org.uk/learn/infodocs/peo ple/pp-einstein.html.

[4] Ronald W. Clark, *Einstein: The Life and Times*, New York, Avon Books, 1972.

[5] Sarah Chankin-Gould, *Preventing Nuclear Proliferation in Latin America: The Treaty of Tlatelolco*, FAS Public Interest Report, Winter 2004, Volume 57, Number 1.

[6] Enrique Román-Morey, Carnegie International Non-Proliferation Conference, Washington D.C. March 16–17, 2000.

[7] Tariq Rauf and Fiona Simpson, Arms Control Today (online), December 2004.

[8] *Third World War Menace Looming*, Pravda, 17 January 2005.

[9] Elise Boulding, *Cultures of Peace: The Hidden Side of History*, Syracuse University Press, New York, 2000, chapter 3.

[10] Ibid., chapter 11.

Ana Maria Cetto (IAEO), Luis de la Peña Averbach

Ana Maria Cetto was born in 1946. She was the former head of the Department of Physics and Dean of the Faculty of Sciences, and of the Department of Theoretical Physics of the Institute of Physics, UNAM (Universidad Nacional Autónoma de México). She is the Secretary-General of the International Council for Science (ICSU) and member of a number of international scientific organizations. At present, she is Deputy Director General of the International Atomic Energy Agency (IAEA) and Head of the Program for Technical Cooperation, Vienna, Austria.

Luis de la Peña Averbach is her husband and was formerly the President of the Mexican National Society of Physics.

From Nucleus to Nuclear Targeting and Nuclear Proliferation

Alla Yaroshinskaya

The history of the struggle for superiority in the creation of the nuclear bomb extends over 6 key years – starting with the 1939 fascist invasion in Europe to 1945 when the USA carried out its first nuclear test. It is an insignificantly small interval of time even when compared with a human life. By historical measures it is merely an instant. But this instant has transformed, and continues to transform, human existence into a permanent struggle for survival in a nuclear world. Today the world is again on the brink of nuclear war.

And it all began absolutely informidably – with the Renaissance of physics in the beginning of the 20th century. The cascade of brilliant fundamental discoveries on the boundary of the centuries became a harbinger of humanity's transition to a new era, a new attitude and a New World outlook. I shall name only some of these discoveries: V.K. Roentgen's discovery of X-rays (the Nobel Prize in 1901), the discovery of polonium, radium and uranium's natural radioactivity by A. Bequerel, P. Curie and M. Sklodowska-Curie (the Nobel Prize in 1903), the discovery of the electron by D. D. Thomson (Nobel Prize in 1906), the creation by Einstein of the general and special theories of relativity and the formulation of the law of mutual connection between mass and energy that became the base of all nuclear physics (the Nobel Prize in 1921), and the creation of the quantum model of the atom by N. Bohr that opened a new round in development of the nuclear theory (the Nobel Prize in 1922).

In 1911, an Englishman, E. Rutherford, opened a nucleus, and since then the term "nucleus" has been one of the basic concepts of modern physics. In 1914, E. Rutherford opened a proton in a nucleus. In only five years, he managed to carry out the world's first nuclear reaction – transforming nitrogen into oxygen.

Einstein – Peace Now! Reiner Braun and David Krieger (Eds.)
Copyright © 2005 WILEY-VCH Verlag GmbH & Co. KGaA, Weinheim
ISBN 3-527-40604-2

In truth, the first 10 years of the 20th century became the Golden Age of physics. (In Russia this same period has gone down in history as the silver age of poetry.) But this Golden Age of physics was really marked by discoveries that have had far-reaching, fatal consequences – down to Hiroshima and Nagasaki.

A decade of calm then followed. But the beginning of the 1930s brought a new wave of discoveries. In 1932, the Englishman D. Chadwick opened a neutron (the Nobel Prize in 1932). The first nuclear cyclotron was constructed by E. O. Lawrence and M. S. Livingston (the Nobel Prize in 1939). Active research on the nucleus with the use of a stream of particles began. The French scientists Frederik and Irene Joliot-Curie bombarded the nucleus by alpha particles, and the Italian scientist Enrico Fermi subjected it to neutron attack. And so began man-made, artificial radioactivity.

The scientific world, not yet suspecting any sinister outcome, quickly moved toward the creation of a nuclear bomb. It took only ten years.

The discovery of nuclear fission of uranium became the transition to a new, more sophisticated level in the development of nuclear physics. The German scientist, a woman, Ida Noddak, was the first who guessed what happens with a nucleus of uranium during its bombardment by slow neutrons. This occurred in 1934. I emphasize it was in 1934 in Germany. This is an important circumstance – by then, Germany had been seized by fascist madness. It is distinctly fearful what would have happened had Ida Noddak's work been understood by her German colleagues. Hitler surely would have had a nuclear bomb before anyone else, possibly even available by the beginning of his attack on Poland on September 1, 1939. But, fortunately, Ida Noddak's colleagues did not react at all to her communications.

Only four years later, in 1938, the German scientist Otto Hahn, who in due time addressed Noddak's discoveries, together with colleagues came to the same conclusion: the nucleus of uranium collapses under bombardment by neutrons. Moreover, it happens in an explosive manner; its particles scatter with huge speed, and the piece of uranium is heated up. This is nuclear energy.

Only after the publication in January 1939 of articles by Hahn and his colleagues, did many other physicists around the world understand what was really happening, and for the first time they discussed the possible creation of a nuclear bomb. By then, the physicist Leo Szilard had already written that this possibility was "extremely dangerous to all humanity."

Somewhere in this period, the free exchange of information ceased among nuclear physicists. Many of them, understanding the global consequences of the development of nuclear weapons, especially during that time, voluntarily engaged in self-censorship. It was extremely important that Hitler would not have the first nuclear bomb. It was clearly apparent that nuclear weapons were now inevitable – it was just a matter of time.

And what was occurring in Russia at the beginning of the 20th century, a country now the world's second nuclear power? In the beginning of the 20th century, Russia had no time for nuclear physics because a succession of revolutions and wars was taking place: the bourgeois revolution in 1905 and the socialist revolution in 1917; then World War I. The Bolsheviks called for transforming this war into a civil war, and they succeeded. (Long before Hitler, Lenin constructed across all of Russia concentration camps in which the Bolsheviks let their opposition rot. In the civil war begun by Lenin, ten million people died.)

During this upheaval, all of Russia's best minds emigrated abroad, mostly to France. Those who disagreed with the new ideology but had not emigrated were deported, while others were put into prisons and camps. It is no wonder that at that time there had not yet appeared any Russian national physics schools, or Russian discoveries in nuclear physics. Russian scientists had no communication with the West, as Stalin's iron curtain now kept our country apart from the global community.

There were just a few exceptions to this: the establishment of The Radium Institute in 1922 by V. Vernadski (he had been given one gram of radium for his research). And several Russian scientists were given permission to go to the West for training. Peter Kapitsa worked for 13 years at the Rutherford laboratory. One of the future fathers of the Russian nuclear bomb J. Hariton also trained there for two years. And the outstanding Russian physicists, L. D. Landau and G. A. Gamov became the pupils of Niels Bohr, while Rontgen's laboratory at Munich University became the school for A. D. Ioffe.

The Soviets' own physics school began to be outlined only in the 1930s under I. V. Kurchatov's management. He headed the Physical and Technical Institute, and he began research on the nucleus only in 1932. The USSR's first physics and chemical scientific magazines began to appear at that time, also.

During this same period, many physicists, especially Jewish physicists, left Germany. They left because of ethnic or ideological persecution by the Nazi regime. Many of them arrived in England and the USA. Some of the best scientists were transported to England by Frederik Lindemann, a personal friend of Churchill. These scientists then became part of the English nuclear program. The same was done also by emissaries of the USA.

In 1938, in England, for the first time in the world, governmental support of and control over nuclear research was established for the military goal of creating a superweapon – a nuclear bomb. A special committee was created at the Aircraft Ministry for this purpose. Thus began the British nuclear program, but England did not become the first nuclear state, nor did Germany.

It is interesting that in April 1940, German and Austrian scientist-emigrants R. Pierls and O. Frisch first presented to the head of the Committee, Henry Tizard, a memorandum, "About Creation of the Superbomb Based on Nuclear Chain Reaction in Uranium." They began a campaign to convince the English authorities of the urgent necessity to start working on creating a nuclear bomb. As a result, their offer was accepted, and soon after, a secret special Committee was created to develop the first British nuclear bomb. By then, a factory to enrich uranium had already been approved as a new project. Britain concentrated its efforts to become the first nation with a nuclear bomb, in order to decide the outcome of the war. Today, it is possible to draw a conclusion: in 1941, Hitler's Germany was probably the first nuclear target of Great Britain's probable first nuclear bomb.

But history took another turn. In 1941, the American Banbridge took part in the confidential meeting of Great Britain's Committee on the nuclear bomb, and he immediately understood everything. His participation resulted in the Committee deciding to cooperate with the USA in a comprehensive expansion in the nuclear sphere. That decision resulted in the subsequent loss of Britain's nuclear weapons program, although the British did not yet realize it, and even created their own state nuclear program.

In 1941, a copy of the British Committee materials casually lay on a table of Vannevar Bush, the head of the US department of scientific research and development. The fall of that year saw the start of a number of agreements adopted between Britain and the USA about information exchange in the world of nuclear research. Further, the British project of industrial uranium production was rapidly transferred to the Americans.

In 1942, Great Britain accepted the offer of the USA to work on development of the weapon together. In 1943, Roosevelt and Churchill signed Kvebek's agreement that practically closed down the British nuclear project. By this time, the Americans had already "pumped out" from Britain everything that was possible, and the British were no longer necessary for them. The British were not even informed of the start-up of the first nuclear reactor in Chicago on December 2, 1942, or the beginning of the construction of the factories in Oak Ridge and Hanford. The door had slammed shut for the English scientists. Jumping forward a few years, I should note that it was only in 1947 that the British accepted the new nuclear program and then spent five years to acquire its own nuclear bomb.

Meanwhile, German scientist-emigrants in the USA were very concerned about Hitler's regime and his preparation for war. Fearing Hitler's ability to create the first nuclear bomb, they sent President Roosevelt a letter along with an appendix on the possible development of the new superweapon. The letter was signed by A. Einstein, though he was a pacifist for all his life. The Uranium committee was soon created, but no action was yet taken.

Then, in 1940, physicists appealed to Roosevelt repeatedly. By then, the occupation of Europe by Hitler was fully underway. But only on December 6, 1941, after an attack of Germany on the USSR, did the White House agree to start a nuclear program. The next day, Japan attacked Pearl Harbor. On August 13, 1942, the administration of the USA ratified the "Manhattan Project." Colonel Leslie Groves became its head and five days later he was made a general.

After three years of enormously intensive work, the construction of factories, and the creation of the first test site, the first two samples of bombs – uranium and plutonium – were achieved. On March 25, 1945, the same scientists Einstein and Szilard wrote to Roosevelt to prevent the production of a nuclear bomb, but they received no answer. On April 12, 1945, President Roosevelt died without having given any orders about the nearly ready nuclear bomb. (There are opinions that shortly before his death, Roosevelt considered the possibility of a nuclear attack against the Japanese fleet.) The Minister of Foreign Affairs of the USSR, Alexey Gromyko, in the book, *The Memorable* (Moscow, 1988, page 294), writes that about this time, Einstein said: "If I had known Hitler did not have a nuclear bomb, I would not have begun to support the American nuclear project." War with Hitler then came to an end.

Attempts to stop the use of nuclear weapons were undertaken also by other scientists. But the train had already left the station. The new US president, Truman, was totally unaware of The Manhattan Project, and was surprised when he learned of the nuclear bomb. On July 16, 1945 at 5:30 am, the first US nuclear test, "Trinity," was successfully carried out.

At the same time, the Potsdam conference of the leaders of the countries of the antifascist coalition was held. On July 24, after the ending of that day's session, Truman came to Stalin and informed him of the USA test of the new weapon surpassing any another had been successful. Stalin, according to all experts, did not even turn a hair. Truman recalled Stalin congratulated him and wished the new weapon "be used against Japan." Did Stalin really state the target for a new American nuclear bomb? On May 31, 1945, the special commission of the USA recommended to Truman the use of the new weapon against Japan, having chosen a target in an area of buildings that could be destroyed easily. It was a terrible, inhumane recommendation!

The members of the commission included five politicians, three military policy scientists, and four nuclear physicists – R. Oppenheimer, E. Fermi, A. Compton and E. Lawrence. Among them, only Fermi was not an American.

What was occurring in the USSR at this time in physics, this period of bloody war against fascism, having lost 27 million citizens over four years? As I have already mentioned, until the 1930s, in Russia there was no nuclear school, there only were separate attempts of separate physicists to deal with the problems of the nucleus. Besides, in the second half of the 1930s, political reprisals began. The outstanding theoretical physicists U. A. Krutkov, and P. I. Lukirsky had been arrested and sent away to camps. On the charge of espionage for the benefit of Germany, the well-known physicist L. D. Landau and others had been arrested. But it was not really the truth. In addition, the Bolsheviks tried to ideologize science. For example, genetics was declared a pseudoscience. And something similar occurred with nuclear physics. Theoretical work on uranium had not been forbidden, but as the well-known nuclearist Y. B. Zeldovich recalls, scientists were engaged in it only in the evenings.

All this slowed development of scientific ideas in the nuclear sphere. But beginning in 1933, in the USSR All-Union conferences on nuclear physics started to be carried out on a regular basis. There was intensive work on nuclear particles. It is important to note also that in 1939 (already after Gan and his colleagues), U. B. Hariton and Y. B. Zeldovich published two articles on the chain reaction of uranium under influence of slow neutrons. (I believe Hariton's experience in laboratories of Rezerford during his training there helped them to do that.)

In July 1940 in the USSR for the first time the question on expansion of work on nuclear issues was included for discussion in the Academy of Science, and the commission on uranium was started. The decision to begin uranium geological prospecting in Central Asia was also made.

But with the beginning of the war with Hitler on June 22, 1941, all nuclear research stopped, and all efforts of scientists were directed to the development of conventional armaments that were needed at the front.

In 1939 the head of the most terrible departments in the USSR – the People's Commissariat of Internal Affairs (the future KGB) – L. Beriya had withdrawn from abroad some soviet secret-service agents who were then killed or sent to prisons and camps. But on the eve of the war with Hitler, the People's Commissariat of Internal Affairs under urgent orders prepared and sent out new secret-service agents, including to London.

One of the outstanding Soviet "nuclear" spies was Vladimir Barkovsky. He arrived in London by way of Japan, the Hawaiian islands and the USA mainland in the beginning of February 1941. Europe already lay at Hitler's feet. Barkovsky's mission was to find and send to Stalin English nuclear bomb secrets. It was very timely – the British had just actively started to work on the bomb. "London residences became the pioneer in the delivery of such information," – Vladimir Barkovsky recalled in his memoirs. (Vladimir Chikov. "The bomb, stolen from the safe". "The Russian Newspaper". 2001, August 22). In September 1941, the Soviet secret service in London broadcast to Moscow the ciphered message that Great Britain had created a Special Committee for the development and creation of a nuclear bomb. The second message from London described a plant for the division of isotopes of uranium and calculations of critical weight of uranium-235. Finally, the full report of the British nuclear activities appeared in Moscow.

Since 1941, the German physicist Klaus Fuchs who emigrated in 1933 to England and later worked two years in the USA at Los Alamos, worked for the KGB. It is known that in 1949 he was "suspected" by the English secret service and kept under surveillance for 14 years for espionage, on the belief that he was a communist. The same is also true for John Kernkross, an employee of a military department of the USA. Bruno Pontekorvo, an emigrant from Italy, the well-known employee of Enrico Fermi, also cooperated with the Soviet KGB. Frequently, they transferred secret information to the Soviet special service under their own initiative. As Soviet agent Barkovsky says, all his Western agents worked for the USSR not for money, but for the cause of communism.

It is a known fact that after the arrest of Fuchs, there followed the persecutions of many US people by the House Un-American Activities Committee. Victims also included the well-known physicist Oppenheimer because he was alleged share communism's ideas.

But until 1943, all invaluable materials passed by Western physicists were kept in the cabinets of the Kremlin and Lubyanka (People's Commissariat of Internal Affairs). No scientist was allowed to see them. Some two thousand pages of special scientific material concerning the manufacture of a uranium nuclear bomb were collected.

Several months prior to the first US nuclear test, the physicist Niels Bohr wrote to Prime Minister Churchill of Great Britain advising him to share nuclear bomb secrets with Russia. He believed that unilateral control over nuclear weapons by a single country would lead to heavy consequences for all humanity. Churchill did not listen to him. Bohr's letter to Roosevelt with a similar warning also went unheeded.

In 1942, Soviet nuclear physicist G. N. Flerov, who was at the front, wrote a letter to Stalin about the necessity of developing a nuclear bomb, and explained in non-technical language its design and force.

Before the war, Flerov had already discovered independently from western scientists the phenomenon of spontaneous nuclear fission of uranium-235. He knew what he was talking about. His letter reached Stalin at the same time as Beriya's report on England's and America's nuclear-bomb programs.

Soviet agent Barkovsky rightly notes that the fault of the USSR's slowness in acting to create a nuclear bomb lies with the chief of the People's Commissariat of Internal Affairs, L. Beriya. If he had considered reports from the Soviet agents more seriously earlier, and had informed Stalin, despite the horrific war, the USSR could have created a bomb earlier than the USA. The question for me is what would have happened in this case? Would Stalin have bombed Hitler and Japan? And perhaps the USA as well? What country would have become the first nuclear target of the bloody dictator Stalin? I do not have any doubts that Stalin would have used a nuclear bomb against Hitler in order to rapidly end the war.

At last, on February 11, 1943, Stalin signed the document creating the Soviet nuclear bomb program. The secret program, headed by a young nuclear physicist, I. V. Kurchatov, was established. He was the first Soviet scientist to whom all classified documents of the Soviet Intelligent Service were shown. He discovered in the documents two things absolutely unknown to Soviet science: First, a nuclear reactor can work using not only heavy water, but also using graphite. Second, plutonium can be used to make a nuclear bomb and a plutonium bomb requires much less critical weight than a uranium bomb.

There were so many and varied documents from the Soviet agents (studied at the Kremlin and by the KGB) that even if Kurchatov had been a superman, it was beyond his ability to become an expert on all the information. Therefore, despite Beriya's resistance, additional physicists were brought onto the project, including Ioffe and Hariton. They had no authority to reveal the KGB's secrets and therefore they had to claim the scientific data as their own discoveries. This created around them an aura of geniuses.

By the end of 1944, a single ingot of one kilogram of pure uranium had been created. In 1945, instead of V. M. Molotov, L. Beriya, the head of the KGB, had been appointed the curator of the nuclear project. Millions of Soviet prisoners worked in total secrecy in uranium mines, knowing nothing about what they actually were doing.

The USSR government was rushing its nuclear bomb program, especially after the first test in the USA. The big question was already about US nuclear ambitions rather than about Germany's. There was no doubt in Moscow that the USA wanted to destroy the Soviet Union with the help of the new weapon. After Hiroshima and Nagasaki, it became even more obvious. Therefore, Stalin set a strong task for the country's scientists – to create a nuclear bomb by 1948.

Beginning in the summer of 1945, the US military-industrial complex began to develop a plan of nuclear attack on the USSR, to define the targets. The first project was named "Strategic Vulnerability of Russia for Limited Air Attack" and was dated November 1945. In 1948–1949, a detailed plan of bombardment of the USSR was prepared. It was supposed to demolish 70 cities and industrial centers, about two thousand subjects in total. It is estimated that up to 2.7 million people would be killed initially, and 4 million more wounded over a month.

Fearing a nuclear attack by the US, the Soviets began to bluff. In 1947, the Minister of Foreign Affairs Molotov (by the way he was one of the most bloody Bolsheviks who sent to camps thousands and thousands of innocent people) declared that "the secret of the nuclear bomb in the USSR is discovered", and that the USSR possessed a nuclear bomb, even though at that time two years remained before the USSR's first test. Some Russian experts say it was a "holy lie" because in this way Moscow tried to protect itself from American nuclear attack. At that time in the USSR in different areas of the country, huge non-nuclear explosions were done to show to the US visible "proof" of the USSR's nuclear testing.

The first real test of a Soviet nuclear bomb took place at the Semipalatinsk test site in Kazakhstan on August 29, 1949. The USSR did not officially declare this test because Stalin was afraid that the USA would carry out pre-emptive strikes on the Soviet nuclear plants. After that, Fuchs was arrested and Truman announced the creation of the hydrogen bomb.

So the first stage of the nuclear era ended and the second one began – a mad nuclear arms race, mutual targeting, and proliferation of nuclear weapons.

Humanity needed 25 years to get to the Nuclear Non-Proliferation Treaty (NPT), to try to push the nuclear genie back in its bottle. But the nuclear monster still does not wish to get back into the bottle even after 35 years since the Treaty's signing. Over the last few years, we have witnessed just the opposite: the world now faces an increased threat of nuclear war.

Even after the reductions of nuclear weapons by both the USA and Russia according to the START-1 agreement, we have more nuclear weapons now than before signing of the NPT. No country publicly reveals exact figures, but according to western experts published in the magazine "Nuclear Proliferation", in 2002 Russia had 5858 strategic warheads and the USA 7013. According to that source, Russia had 4000 tactical warheads, the USA 1620. Russia had in storage 9421 warheads and the USA about 5000. However, do not forget about the nuclear weapons of Great Britain, France and China.

In addition, the general stocks of plutonium are estimated by experts for Russia to be 150 tons, and for USA 99.5 tons. This is enough plutonium to make 40,000 warheads. The stocks of highly enriched uranium are also enormous. There are more than 1500 tons in Russia and 944 tons in the USA, which is equivalent to more than 100,000 warheads.

I also want to remind you that during the Cold War, the highest levels of the USSR and the USA nuclear arsenals were respectively 32,000 and 40,000 units. The scientist Arjun Makhijani, the President of the International Institute for Energy and Environment (USA) gave the fact in one of his articles that in the 1950s the USA planned to use 750 nuclear bombs against Russia. US documents in 1954–55 on nuclear war with the USSR indicate this number of nuclear bombs would be enough to transform Russia "into smoking radioactive ruins within two hours."

These are worrying numbers.

Now I will talk about other countries. It is known that India and Pakistan have declared themselves to be nuclear powers (they did not sign the NPT) and are teetering on the brink of regional nuclear conflict. The nuclear potential of Israel is also not a secret. Further, 36–44 states have nuclear reactors in nuclear power plants and in research facilities. That is why, according to the Comprehensive Test Ban Treaty, they possess the technical ability to make nuclear weapons. According to a Russian Secret Service investigation published in 1995, 20 countries (!) are on the way to becoming nuclear weapon states.

In 1995, at the conference on the prolongation of the NPT where I worked in the official Russian governmental delegation as an Advisor to the Russian President, it was clearly shown that a nuclear apartheid exists in the world. It looks likes there are two ways of overcoming this: constructive – to achieve implementation of Article VI of the NPT- and destructive – to overturn the NPT and to arm ALL countries with nuclear warheads.

It seems the second scenario is developing at a fast tempo and sooner or later it will lead to nuclear war. The statements of one of the most high-ranking international officials is interesting on this point. On January 26, 2004, the Director General of the International Atomic Energy Agency (IAEA), Mohammed El Baradei declared that "danger of nuclear war was never greater than now." And further: "The Nuclear Non-Proliferation Treaty does not interfere with a state engaged in producing enriched uranium or even buying nuclear materials that can be used for military purposes. If any of these 35–40 countries currently a signatory of the Nuclear Non-Proliferation Treaty decides to leave it, it can create nuclear weapons within months." It is the truth. I wrote about this in my articles ten years ago. Both then and today, the greatest obstacle to implementing the Nuclear Non-Proliferation Treaty is the illegal sale (or confidential sale, on the basis of secret bilateral contracts) of nuclear technologies.

Here is an overview of the largest leakages of nuclear materials during the years of the NPT's existence:

In 1969, almost right after the signing of the NPT a German cargo ship set sail from Antwerp with 200 tons of uranium. The ship's documents show that the uranium was intended for an Italian chemical firm. But that cargo ship did not reach its port of destination, Genoa. Many months later it was located in a Turkish port loaded with other cargo. The international nuclear services has no information about the missing uranium. Only a year later, one of the CIA's employees said 561 barrels of uranium was sold to Israel. This amount is enough to produce weapons-grade plutonium for 33 small nuclear bombs.

In the middle of the 1970s, in the USA, 4 tons of enriched uranium and plutonium mysteriously disappeared.

In 1978, the British Department of Atomic Energy's inventory shows that during 1971–1977 100 kilogram of plutonium disappeared from atomic power stations in Great Britain.

It is relevant to recall that China tested a nuclear bomb in 1964, and a thermonuclear weapon in 1968. India conducted nuclear tests in 1974. A nuclear bomb in Pakistan became public knowledge in 1984. It has conducted tests for 4 years. Sweden possessed all the resources to make a nuclear bomb around 1957, but it signed the Nuclear Non-Proliferation Treaty in 1968. The Republic of South Africa had six nuclear warheads in 1989, but in 1991 it signed the NPT.

Most of these countries could not independently develop nuclear weapons or make enough uranium and plutonium as fast as they have acquired these materials. From this, we can conclude clearly that they bought all the necessary elements – from the engineering specifications to fissile materials.

The truth about the nuclear black market is much more terrible than the frightening newspapers and TV accounts. They say that we are being threatened by nuclear terrorists. I agree, because we are being threatened, most of all, by a system of secret bilateral state contracts dealing with nuclear technologies. I have absolutely no doubt of it. Here are the facts:

In 1945–1946, the USA gave information on a bomb to Sweden.

In 1970, the USA transferred nuclear materials and technologies to Israel.

In 1974, the USA, France and Israel shared nuclear secrets with the Republic of South Africa.

In 2003, Russia began to build a nuclear power plant in Iran. In the last year under pressure of the IAEA, Iran has admitted that it did receive nuclear technologies, including a centrifuge for enriching uranium from Pakistan.

In 1956–1960, within the framework of the Soviet and Chinese program to transfer nuclear rocket technologies and to train Chinese in high schools of the USSR, nuclear secrets appeared in the Chinese People's Republic.

China has sold nuclear secrets to India, Pakistan and North Korea.

Subsequently, a chain reaction was created by these activities:

Pakistan has shared nuclear knowledge with Libya. Pakistan plans to sell a bomb to Saudi Arabia. In October 2003, the *New York Post* referring to Aaron Zeevi, the representative of the Israel Army, Major General, stated that Saudi Arabia was negotiating with Pakistan for the purchase of nuclear warheads.

Argentina, Brazil and Syria are probably on the way to creating nuclear bombs.

And certainly it is more likely in these conditions that nuclear terrorists can more easily take advantage of the increasing chaos in the world of underground nuclear trading to achieve regional, if not global, nuclear proliferation.

Today, China and Israel have the technological capacity to produce mini-nukes. In addition, according to the Stockholm International Peace Institute China possesses nuclear landmines and Israel has explosive nuclear devices. But there is no information about sizes, weight and nuclear power of these.

There is no guarantee that an action of a nuclear terrorist will not provoke a chain reaction at a national level. In this, the Middle East is especially potentially dangerous. And I always await news from there with fear.

The years after the USSR dissolved were characterized by deep stagnation where the nuclear non-proliferation regime was concerned. The situation in Russia (and in the world) has improved little with the coming to power of Vladimir Putin in Russia. The Russian Duma immediately ratified the START-2 and CTBT under his pressure, and this has somewhat improved the international nuclear climate. Then presidents George W. Bush and Vladimir Putin signed the Strategic Offensive Reduction Treaty (SORT) to reduce the nuclear arsenals of the two states to 2200 warheads each over ten years. However, this Treaty is designed to placate a concerned world rather than to be an effective step to peace.

During this same period, the USA has taken the following steps that have practically broken down the efforts of the international community to stop nuclear proliferation and a new arms race: walked away from the Anti-Ballistic Missile Treaty of 1972; declared a new nuclear doctrine that allows use of nuclear weapon in local conflicts; reanimating the so-called Star Wars program; constructed national and local missile defenses using space. All these actions only serve to provoke the non-nuclear countries to become nuclear nations by any way possible.

The published parts of the 2002 Nuclear Posture Review (NPR) of the USA also do not add optimism. Russia and other countries indicated there are likely potential nuclear targets of the USA.

Actually, the Russian establishment was shocked by this news from NPR on the targeting of Russia because on January 14, 1994, Presidents Yeltsin and Clinton signed the Moscow Declaration on mutual non-targeting. (Russia has also signed one with China.)

The experts suppose that the targeting of Russia is the reason why US President Bush could not agree with Putin's proposal to cut down the nuclear arsenals of both countries to 1,500 units. And although Bush could offer no explanation, he repeated many times that Russia and the USA are not enemies. It is also clear that the nuclear arsenals of Britain, France and China offer no reason to worry. So, why not reduce the number of nuclear weapons even further if Russia agrees to do that?

The reason is that although President Bush says that "the preconditions of a choice of the purposes for the nuclear weapon in the days of the Cold War no longer dictate the size of our arsenal," they actually do. Today, as well as in days of the Cold War, the threat from Russia defines nuclear arms planning in the USA. Nuclear strategists of the USA continue to believe that for successful deterrence, the United States needs to have an opportunity to limit damage in case of nuclear war. It could even be done pre-emptively, on 2,200 targets in Russia, including 1,100 nuclear arms targets, 160 control centers, 500 conventional arms targets, and 500 military industrial-enterprise targets.

I think this number of targets will be reduced if Russia cuts back its nuclear arms. However, on the US side, Bush will not reduce US nuclear arsenals to 1,500 units because the US still wants a large number of targets in Russia.

It is interesting to know how Russians view the issue of de-targeting Russian nuclear weapons. According to a 1999 sociological poll, the majority support de-targeting. Further, only 12% believe that the weapons should be kept in storage, while 82% maintain that missile systems must be kept on full military alert.

It is estimated that, at present, Russia would be able to launch approximately 2,100 nuclear warheads within a few minutes of the command being given; even with the full implementation of START 1 and 2, this figure would still amount to several hundred.

There is still fear among the Russians that they might be the target of nuclear attack from another state. 52% of the poll respondents consider such an attack to be possible, while 38% think it impossible. The residents of Moscow and St Petersburg – the most likely targets of nuclear strike – are relatively sure that there is no threat of nuclear attack (62%, against 28% who consider it possible). Considerably more worrisome is the possibility of nuclear facilities in Russia becoming the target of terrorists. 89% of those questioned believe that terrorists may attack such facilities, while only 7% believe in the security of Russian nuclear facilities that they consider terrorist attacks to be impossible. Strikingly, only 5% of the elderly believe the facilities to be secure.

I do not support sole reliance on de-targeting. For me, targeting is like a lamp. If you switch it off, it does not mean that the lamp will not work. It takes only a second to switch on it again. The same is true with de-targeting. The states involved need only several minutes to retarget their nuclear weapons.

Since the fall of the Berlin wall in 1989, military circles in the USA have continuously created doctrines and justifications for using nuclear weapons against those countries that they believe have or are developing weapons of mass destruction. The Joint Chiefs of Staff have put forward the new nuclear doctrine that allows use of nuclear weapons in regional conflicts. The US strategic command, Air Force and Navy have modernized strategic reconnaissance and nuclear arms to more effectively strike targets at any point of the globe.

The US Stratetigic Command (STRATCOM) apparently develops lists of the prospective targets that will be transferred to regional commands of US military forces. Moreover, under the new concepts that are put forward by the military, nuclear research centers are developing new types of nuclear weapons of low power – so-called "mini-nukes."

STRATCOM creates confidential lists of targets, named "silver books." Such "silver books" could be transferred to the European, Atlantic, Pacific and Central commands. According to publications, the first "silver book" is already prepared for commanders of armies of the USA in Europe. An unnamed, high-ranking officer familiar with this concept informed *Jane's Defence Weekly* in January 1995, that "the various variants of actions in relation to the countries, the organizations or groups which are the serious threat for proliferation will be collected in the silver book. Strategic Command is going to create the list of targets and full range of weapons and systems of delivery that can strike every target by nuclear or conventional weapon."

Russia is concerned about an expansion of NATO and the US military into territories of the former Soviet Asian and Caucasus republics. For example, Uzbekistan and Georgia have already announced that they are strategic US and NATO partners and US military bases have already been established there. Turkey prepared an airport according to NATO military standards in Marneuli, Georgia, which can handle many types of aircraft, including heavy bombers. According to Russian military experts, the airport restructuring may be related to the US missile defense system and may be used as a place to deploy anti-missile laser weapons systems. These kinds of laser weapons already exist in the United States.

The Bush administration claims its Nuclear Posture Review is directed against so-called rogue nations, but geopolitically it will also provide the United States with the capability to control Russian territory. If the US deploys a Boeing 747 with laser weapons on the territories of Georgia, Kyrgyzstan, Kazakhstan or Afghanistan, it will be able to control not only Iran, Pakistan and parts of India, but also parts of China and Russia.

Do leaders of the USA and the NATO believe these countries will agree with such a situation near their borders? It is obvious, that such developments will give them a serious reason to take measures for their own safety. Very probably they will respond to it by developing and producing new kinds of weapons, including nuclear.

And who will seriously believe that China will resignedly agree that the USA will "cover" Taiwan and Japan by a nuclear "umbrella"?

After the publication of part of the NPR in March 2002, the Minister of the Foreign Affairs of the Russian Federation, Sergey Ivanov, declared Russia is not going to destroy some parts of dismantled warheads.

And President Putin declared radical modernization of the Russian nuclear forces. This modernization already has begun. Russia has now also declared that it will employ a nuclear first-strike if there is a threat to national safety. In the recent report by the Minister of Defence of Russia, "Actual Questions of Development of Armed Forces of Russia," it states that Russia gives active fighting status to nuclear weapons, and it also does not exclude using pre-emptive strikes.

Flights of NATO aircraft over the former Soviet countries of the Baltic – new NATO members – have given a new boost to the Air Forces of Russia. In March 2004, Minister of Defence Sergey Ivanov declared the creation of a new Russian aerospace defense. Work on radical modernization flight systems of Theater Missile Defense (TMB), and the ground automated complex for management of space apparatus have already begun.

Soon after the nuclear destruction of Hiroshima and Nagasaki, American scientists founded a monthly journal, *Bulletin of Atomic Scientists*. On its cover they placed a clock named "doomsday clock" whose hands show ten minutes to midnight.

Since 2002, they have stopped the hand of doomsday clock at seven minutes to midnight. Probably, they are bigger optimists than I am. It seems to me that with India and Pakistan declaring the possibility of an exchange of nuclear strikes in 2002, the US declaring the probable use of nuclear weapons in local wars, and after that Russia declaring the possibility of a pre-emptive nuclear attack, the hand of the nuclear clock should be at only one minute before midnight, nuclear amargeddon.

Now at last, humanity should seriously reflect on how to turn this hand back from the brink of a nuclear precipice.

Alla Yaroshinskaya

Alla Yaroshinskaya, born 1953, graduated in journalism. She was a member of the Supreme Soviet during the Gorbachev period and an advisor of President Yeltsin. In 1992 she was awarded the "Alternative Nobel Prize". She is the President of the Ecological Charity Fund and Co-Chair of the Russian Ecological Congress.

Part 3
Striving for Peace

"Striving for peace and preparing for war are incompatible with each other, and in our time more so than ever."

Albert Einstein

Einstein – Peace Now! Reiner Braun and David Krieger (Eds.)
Copyright © 2005 WILEY-VCH Verlag GmbH & Co. KGaA, Weinheim
ISBN 3-527-40604-2

Thoughts on Conflict Resolution in the Tradition of Albert Einstein

Oscar Arias Sanchez

How ironic it is that those who work for peace are so often accused of acting in a naïve or romantic manner, while the proponents of violence justify their stance under the guise of historic inevitability! Of the ninety three organizations and individuals that have received the Nobel Peace Prize, surely the vast majority, including myself, have faced criticisms that our work was nothing but an attempt to impose an irresponsible idealism on a world that responds only to force. But if one truly believes, as Miguel deCervantes said, that peace is the greatest good that men can hope for on this earth, then one will quickly learn how important it is to share in the struggles of kindred spirits who are not afraid to demand a higher morality in the actions of nations and individuals.

Indeed, those who have been able to change the world for the better are most likely to have been like the Man of La Mancha, who charged every windmill he could find and never lost sight of the beauty in life. One of the greatest such Quijotes in modern history was Albert Einstein. In 1931, as the scourge of fascism spread through the heart of Europe, the scientist told an American reporter:

> I am not only a pacifist, but a militant pacifist. I am willing to fight for peace ... Is it not better for a man to die for a cause in which he believes, such as peace, than to suffer for a cause in which he does not believe, such as war?[1]

Rather than see peace activists scorned and persecuted, as had occurred during World War I, or passive and apathetic, as was largely the case during the rise of the Third Reich, Einstein wanted to witness peace, not war, become the true protagonist of human affairs in the twentieth century.

Einstein – Peace Now! Reiner Braun and David Krieger (Eds.)
Copyright © 2005 WILEY-VCH Verlag GmbH & Co. KGaA, Weinheim
ISBN 3-527-40604-2

Instead, during Einstein's lifetime the world saw the rise of the idea of an "armed peace," political stability based on the fear of mutual annihilation. Through all the terrifying developments, the great physicist struggled against the perception that militaries had a monopoly on such values as sacrifice, valor and nobility. From Einstein, we learn that the pursuit of peace, like physics, literature or music, is a great and challenging enterprise that requires creativity, courage and perseverance. It depends on dreams, yes, but also on actions; idealism, but also pragmatism; abstract reasoning, but also detailed knowledge of specific places and personalities.

When I received the Nobel Peace Prize in 1987, I said that peace is a process that never ends, the path forged by many people in many lands. It is an attitude, a way of life. Working in peace requires us to work together, with respect for our differences and an abiding regard for our common interests. It is striking that the pluralistic and tolerant values required of peacemakers are also integral to the field of scientific research. Indeed, Einstein believed that with their constant need to share hypotheses and compare results, regardless of national or cultural borders, scientists are naturally inclined to favor pacifist goals. Therefore, for both the peacemaker and the scientist, a common theme emerges: the absolute necessity of dialog to promote understanding and achieve a long-awaited breakthrough.

Although it is often scorned for being too slow, the ancient tool of dialog always proves to be the most appropriate instrument for resolving conflicts. Dialogue gives all sides the chance to gain new insights and discover unexpected solutions to seemingly intractable problems. However, such breakthroughs take time, and those who have in their hands the lives of their fellow men and women, the tranquility and material well-being of entire societies, do not have the right to give in to exasperation. As a scientist patiently labors to explain a puzzling phenomenon of nature, so too must a political leader have the wisdom not to fall into the trap of frustration. Every conflict that is resolved through violence leaves behind a trail of resentments that are very difficult to heal. Every dispute that is resolved through dialog is a step forward.

Today, conflicts from Chechnya to Colombia prove that reconciliation is a profound and difficult process that involves years of labor, setbacks and perseverance. To believe in the possibility of lasting solutions to these long historical struggles, it is not necessary to believe that negotiations are infallible. We know that parties are often intransigent, that leaders may fail to live up to their obligations and responsibilities, and that violent dissenters can obstruct even the most popular commitments to peace.

Although negotiations may take time and try our patience, it is clear that the alternative is far worse. When pacts are broken, it is more sensible to return to the negotiating table, than to engage in the dangerous gamble of military retaliation. When tensions increase, it is wiser to seek a better understanding of your opponent's position, than to shut him out completely.

Besides allowing parties to avoid violence, another major benefit of dialog is the opportunity to begin to unravel the knots of hatred and set the foundations of tolerance and prosperity for future generations. Cease-fire agreements are important, but any accord for a firm and lasting peace must address the root causes of conflict, which are almost invariably related to issues of poverty and social inequality. It is foolish to ignore the links between poverty and conflict, yet many leaders and analysts treat them as completely separate problems.

Resources for fighting poverty are not lacking in the world. What we have been experiencing, rather, is a lack of will. When entire economies enter into crisis, wealthy governments quickly gather the billions of dollars needed for a bail-out. But in the face of the crises of poverty and hunger, illiteracy and disease that affect so many of our brothers and sisters, where is the rapid response? Where is the will to gather up the resources that could feed and house everyone, that could provide safe drinking water, basic health and sanitation, and at least elementary education to the world's population?

In the past, we have been led to believe that it is unrealistic to think in terms of resolving the world's poverty crisis. But if we take a close look at the priorities of our governments, we will see that the problem is not one of scale, but one of vision. In the year 2004, 956 billion dollars were dedicated to military expenditure worldwide.[2] According to the World Bank, an annual diversion of ten per cent of world military spending would fully fund the Millennium Development Goals, which include eradicating extreme poverty and hunger, achieving universal primary education and drastically reducing child mortality by the year 2015. In other words, it would take only a modest shift in global priorities to resolve humanity's most pressing development challenges.

Einstein said, "We must be prepared to make the same heroic sacrifices for the cause of peace that we make ungrudgingly for the cause of war." Yet, the world's most powerful leaders continue to subscribe to the old Roman adage that whoever wants peace should prepare for battle. By acting on this preposterous assumption, we have produced a world that places priority on buying and peddling weapons over feeding children, thereby aggravating the very cycle of poverty, injustice, and conflict that we seek to break. The problem is that we keep trying to break the cycle in the same way: by fighting wars. Some people seem to think that if we can put down just one more rebellion, if we can bomb just one more country into submission, if we can arm just one more insurgency, then we can concentrate on eliminating poverty. This is a tragic fallacy.

Recent history has shown that countries with powerful militaries tend to use them too quickly, instead of giving diplomacy a real opportunity to avert conflict. At the beginning of the Cold War, Einstein warned that "competitive armament is not a way to prevent war. Every step in this direction brings us nearer to catastrophe... I repeat, armament is no protection against war, but leads inevitably *to* war."[3] While he was clearly referring to the atomic bomb, Einstein could have easily been talking about the effects of conventional weapons such as assault rifles, tanks and helicopters, especially on those countries that suffer not only the violence of war, but also the violence of poverty.

Convinced of the need to attend to the root causes of conflicts, I have been compelled to reject the ferocious logic of militarism and of the global arms trade in particular. My dedication to this cause stems from the painful recent history of Central America. When I assumed the presidency of Costa Rica in 1986, Central America was being torn apart by three civil wars, which had become a proxy battleground for the East–West clash of the Cold War. I learned a fundamental lesson from living through those times: peace cannot take root unless the deepest causes of conflict are brought to light, examined, and publicly discussed. Arms betray this delicate process by adding to intolerance, deepening present grievances and making agreement more distant. This was one of the precepts of the 1987 peace accords signed by five Central American presidents, which among other things demanded the immediate end of arms shipments from the United States and the Soviet Union to Central America.

At the end of my presidency, I was compelled to continue to speak out on the devastating effects of the arms trade, which was clearly contributing to lethal poverty, extreme levels of inequality and overall despair in many parts of the world. I believed that the end of the Cold War provided a significant opportunity not only to halt the massive production and circulation of arms, but also to destroy once and for all the vast arsenals already in existence. But it was not to be. The United Nations has estimated that since 1990, armed conflicts have killed as many as 3.6 million people worldwide and injured many millions more. Particularly tragic is that civilians, not soldiers, are increasingly the victims-accounting for more than ninety per cent of deaths and injuries. Shockingly, at least half of civilian casualties are children.[4]

If for nothing else than the sake of human dignity, we must find a way to ensure that this new century is less bloody than the last. Humanity has advanced so far in the science and the art of making peace through peaceful means that we cannot afford to slip back into the old ways of responding to conflict. Scientific innovation, command of technology and the maturity of ethical and political ideas put a better world within our reach. We should not permit the talent, energy and richness of generations to be squandered on priorities that defy reason.

Therefore, I have for many years supported the adoption of a universal code on arms sales that would ban transfers of weapons to governments that repress fundamental democratic and human rights, or that commit acts of armed international aggression. The first draft of this code, today known as the Arms Trade Treaty, was launched in 1997 by me and seven other Nobel Prize laureates: Ellie Wiesel, Betty Williams, the Dalai Lama, José Ramos-Horta, and representatives of the International Physicians for the Prevention of Nuclear War, the American Friends Service Committee and Amnesty International. To date, over twenty Nobel Prize winners, a growing group of governments and thousands of individuals and organizations have expressed their faith in the Arms Trade Treaty as both morally sound and politically necessary.

Clearly, a campaign to regulate the global arms trade brings us head to head with one of the world's most entrenched interest groups, and it could take years, even decades, to move forward. In this struggle, the moral and political leadership of civil groups, from schools to church councils to public action organizations, is fundamental. Since October of 2003, a grassroots campaign to ratify the Arms Trade Treaty into a binding piece of international law has been advancing in seventy countries around the world, and people have found innovative ways to combine the cause of international arms control with a kaleidoscope of local issues. In Brazil, for instance, the NGO *Viva Rio* has advocated national gun control laws, while building youth clubs and microcredit programs in poor neighborhoods affected by gun violence. And in Costa Rica, the Arias Foundation for Peace and Human Progress has launched a public education campaign on the public health impact of small arms, with a special component for peace training in the public schools.

It has been thrilling to watch in the past decade as the Arms Trade Treaty has gathered worldwide momentum, a rising tide that grows out of the tiny ripples of every individual act of creativity and leadership. Ultimately, this action is based in a firm commitment to building a better future, in which all people can live full and creative lives, free from the restrictions of grueling poverty, of intellectual oppression, of the rule of men with guns. Such a vision is not worth just dreaming for, but fighting for too, with the spark and perseverance of Einstein's "militant pacifism."

References

[1] From an interview during a visit to the United States. Quoted in: Alfred Lief, (Ed.), *The Fight Against War*, John Day, New York, 1933. In: Alice Caprice, (Ed.), *The Expanded Quotable Einstein*, Princeton University Press, Princeton, 2000.

[2] Economists Allied for Arms Reduction, *News Notes*, Sept. 2004, www.ecaar.org/Netwotk/NewsNotes/Sept04.htm.

[3] From a United Nations radio interview, June 16, 1950, recorded in the study of Einstein's home in Princeton. In: Caprice, op. cit., p. 182.

[4] United Nations Development Programme, *Human Development Report 2003: Millennium Development Goals: A Compact among Nations to End Human Poverty*, Oxford University Press, New York, 2003.

Oscar Arias Sanchez

Oscar Arias Sanchez, born 1940, was the President of Costa Rica from 1986 to 1990 and Nobel Peace Prize Laureate in 1987. He holds international status as a spokesman for the developing world. Arias works tirelessly to maintain peace both in Costa Rica and in the wider area of Central America.

The Future of Our World[1]

Ahmed Zewail

Over the last century, our world has experienced at times a "beautiful age" with promises of peace and prosperity, but then some imposing forces changed the entire landscape. History reminds us of recurrences, and the current state of the world is not so different that we may ask – what political and economic forces cause such disorder in a world seeking prosperity through globalization and revolutionary advances in technology? Here we will address the need for a rational world vision that must take into account developments in the population of the have-nots and dialogues of cultures. It is a vision of economic, political, religious, and cultural dimensions in world affairs. Only with such a vision can we shape a bright future for our world.

Excellencies, Ladies and Gentlemen

It is a great honor to give this year's U Thant Distinguished Lecture at the United Nations University in Tokyo. I applaud the purpose of the lectureship named in honor of Mr. U Thant, the Secretary-General of the United Nations from 1961 to 1971 and the man who had the vision to establish this University. I would like to take this opportunity to thank the UNU Rector, Professor van Ginkel, the Director of the Institute of Advanced Studies, Professor Zakri, and the President of the Science Council of Japan, Professor Yoshikawa, for making this event possible with thoughtfulness and style. I also wish

---

1) 5th U Thant Distinguished Lecture Series, United Nations University, Tokyo, April 15, 2003. 1st Lecture given by Prime Minister of Malaysia, Mahathir Mohamad; 2nd by President of the Republic of South Africa, Thabo Mbeki; 3rd by President of the United States, William J. Clinton; 4th by Nobel Peace Prize Winner, Norman E. Borlaug.

Einstein – Peace Now! Reiner Braun and David Krieger (Eds.)
Copyright © 2005 WILEY-VCH Verlag GmbH & Co. KGaA, Weinheim
ISBN 3-527-40604-2

to acknowledge the Ambassador of Egypt Dr. Mahmoud Karem for his warm welcome.

Last year the lecturer in this series, President Bill Clinton, spoke about globalization and our shared future, and the year before Prime Minister Dr. Mahathir Mohamad spoke about globalization, global community, and the UN. Both speakers were concerned about the new emerging world and opportunities for prosperity and global unity. Today I would like to share with you my thoughts about the future of a turbulent world in the present state of economic and political disorder.

The title of my lecture has several implications that I should clarify. It gives the impression that I know the future or the science of futurology. I do not, and in fact I am aware of many predictions made in the past that turned out to be incorrect. What I have in mind is to paint a future that benefits from our history and our rational thinking; a future shaped by *Homo Sapiens* – the species with the greatest brain power on Earth. For this future, I shall present what I envisage for a world of peace and prosperity and how we can achieve our goals with justice and fairness. But first, let me take you inside a time machine to "travel in time" and see what history will tell us.

The Beautiful Age

Nearly a century ago, the world of 1870 to 1914 had an optimistic outlook. The French called the decades before WWI, which broke out in 1914, "The Beautiful Age – La Belle Époque". The world was experiencing the same upbeat spirit of the global community that Mr. Clinton and Dr. Mohamad spoke about here in our present world. Peace and prosperity were on the horizon. The material standard of living was on the rise, democratization was on the rise, continents were being connected by railroads, steam ships, automobiles, airplanes, the telegraph, and the telephone. Man conquered the last uncharted territory of world maps, the North Pole in 1909, the South Pole in 1911, and the United States became the land of promise for millions. Achievements in the sciences, literature, and peace were honored by the awarding of the first Nobel Prize in 1901. (The Nobel Peace Prize centennial publication has indeed given the reasons for calling this period the Beautiful Age.) At that time, the principal force was the force of science and technology that was creating a better life for humankind.

What went wrong then? Great powers were hungry to conquer lands and resources in Africa, Asia, and the Pacific. Control over raw materials and markets and strategic positioning in the world were the driving forces. The power gained by industrialized nations gave them a thirst for the right to rule, and in some cases oppress, those who did not have power. People from other parts of the world could only acquire a Western level of advancement by learning to think like Westerners, and missionaries often defined civilization as a combination of Western religion and science.

The Great Powers formed alliances and Europe was experiencing nationalism. Germany, Austria–Hungary and Italy formed the Triple Alliance, and France, Russia and Britain formed the Triple Entente. The empires of Russia and Austria-Hungary competed for influence over the Balkans following the disintegration of the Ottoman Empire, "Europe's dying man". The First World War began, and the rest is history.

The World of Today

Today, one hundred years later, the analogy may be telling of the dynamics in our present world. In the recent era of globalization (1991–2000), the world looked beautiful again, coming together by the force of global economy and global political ties. The policy of apartheid in South Africa was abandoned, Nelson Mandela was released from prison, and was elected President in 1994. Even the Gulf War of 1991 – strategic to the control of oil resources – appeared to have a moral dimension, namely the return of Kuwait to its people. The solution to the Palestinian–Israeli conflict was on a hopeful track with the signing of the Oslo Accord in 1993. European economic and political cooperation took on a new dimension with the creation of the European Union, and Japan and other so-called Asian Tigers took a major role in world economic developments. United Germany gave the world a new hope for unity and the end of an era – the world of 1946–1963. This world of the Cold War and nuclear armament appeared to have changed into a world of globalization in the 1990s.

Science and technology were again the real forces for achieving the new world status. Information technology brought the world to village-type communities. Advances in the new knowledge of lasers, semiconductors, biotechnology, and the like have transformed our lives with revolutionary improvements in communication and health, and we even began to dream about a future on other planets.

This is not to say that the world now is perfect. Conflicts are still raging in parts of Africa and HIV/AIDS continues to take the lives of large numbers of people in the sub-Saharan countries. Human rights violations and occupation by force continue in the world of globalization. As I speak today, the Iraqi war is taking the lives of innocent people, and the Palestinians are still under occupation. In Europe, there was the horrific ethnic cleansing in the Balkans and the conflict between the Catholics and Protestants in Northern Ireland continues to this day.

Notwithstanding these conflicts and disorders, the nations of the world on the whole are aiming for a united globe through understanding and cooperation – the role of the UN – and through economic developments – the role of globalization. The desire to achieve more peace and stability through global cooperation is articulated, for example, in the Millennium Development Goals (MDGS), a decree endorsed by all member states of the United Nations in September 2000 with the objectives of attacking global problems such as poverty, diseases, and education for all people, from Nairobi to New York. Through cooperation, many agreements and accords have been reached: the disarmament agreement between the United States and Russia known as START (Strategic Arms Reduction Treaty); the peace agreement for a NATO and Russian partnership; the agreement for the banning of landmines; the UN International War Crimes Tribunal; and global conferences to address problems such as the environment, water resources, and AIDS.

World Disorder and Superpower

What then is causing the current disorder? In my view there is a short-term cause and a long-term problem. The September 11, 2001 horrific attack on the United States has caused an impulsive impact on the only superpower in the world. The country has been insulated from external wars throughout its history, geographically distanced by the Atlantic and Pacific Oceans. Moreover, the political system, which is greatly influenced by strong lobbying and capitalistic media, at times has created a gulf between the United States and other countries. The United States is a unique country and the diversity in its population has resulted in an amalgamated culture – a melting pot. But this new culture is not necessarily knowledgeable about the original cultures of its people. The United States also knows it is especially unique and possesses the ultimate power – that of science and technology – and this power makes it a ruling force over world economy, markets, and military status.

America is still in a state of shock and disbelief, and the response in the country varies from moderate to fanatic. Sadly, the September 11th attack may have coincided with extreme religious and political agendas. With this confusion and confluence, now is the time that America needs visionary leadership the most. A vision of unity is in the best interests of all. The United States has a responsibility to lead the globe into becoming a united world. The world still remembers the Marshall Plan and the Peace Corps as examples of uniting initiatives. America cannot afford to alienate people around the world, and it must apply the same standards of fairness at home and abroad. We must all look for the real sources that kindle terrorism and not try to camouflage the real reasons behind it. In the long run, the key is not to ignore the have-nots, not to ignore the frustrated part of the world, politically and economically, and to recognize that poverty and hopelessness are the primary sources of terrorism and the disruption of world order.

The World of the Have-Nots

In our present world, the distribution of wealth is skewed. Only 20% of the population enjoy the benefit of life in the developed world, and the gap between the haves and have-nots continues to increase. According to the World Bank, out of the 6 billion people on Earth, 4.8 billion live in developing countries, 3 billion live on less than $2 a day, and 1.2 billion live on less than $1 a day – an amount that defines the absolute poverty standard. About 1.5 billion people have no access to clean water, with health consequences like waterborne disease, and about 2 billion people are still waiting to benefit from the power of the industrial revolution. The per capita gross domestic product (GDP) in some developed Western countries is $35,000, compared with about $1,000 per year in many developing countries, and significantly less in underdeveloped countries.

This difference in living standards by a factor of 100 or so between the haves and have-nots ultimately creates dissatisfaction, violence, and racial and ethnic conflict. Evidence of such dissatisfaction already exists and we have only to look at the borders of developed and developing or underdeveloped countries; for example, between the United States and Mexico and between Eastern and Western Europe, or between the rich and the poor in any nation. Of similar effect is the frustration caused by double standards in international disputes and in the support of undemocratic or even corrupt regimes for the sake of national economic or political gains.

Some believe that globalization is the solution to problems such as the economic gap, the population explosion, and social disorder. Globalization, in principle, is a hopeful ideal that aspires to help nations prosper and advance through participation in the world market. In practice, however, globalization is better tailored to the prospects of the able and the strong, and, although of value to human competition and progress, it serves only that fraction of the world's population who are able to exploit the market and the available resources.

Nations must be ready to enter through this gate of globalization and such entry has its requirements. Among these requirements are the following: computer and internet literacy, a minimal level of bureaucracy, accessibility to sources of knowledge and information, the entrepreneurial spirit, efficiency in management, and the clear and just applications of the law. With a new system of education and the development of a science base we can hope for an effective globalization. These cannot be achieved without partnership.

World Partnership and World Science

It is clear that world order requires a new and comprehensive partnership between the developed and developing worlds. In my view, science and education are the real glue for binding different cultures and for achieving progress and prosperity. Science is the only fundamental and international language of the world. The developed world is developed because of its scientific and technological power. In this century, knowledge-based societies will capture the lion's share of the world economy and prestige. But, how can the developing world reach such a high state of achievement in science and use the benefits for the betterment of the have-nots? And what will the benefits be to the developed countries?

In the past five years, the scientific community worldwide has published about 3.5 million research papers. Europe's share is 37 per cent. The share of the United States is 34 percent. The Asia/Pacific share is 22 per cent. Other places, representing 70 to 80 per cent of the world's population living largely in the developing world, contributed less than 7 per cent of these scientific articles. Put in a different way, as Mr. Kofi Annan pointed out recently, "Ninety-five per cent of the new science in the world is created in countries comprising only one-fifth of the world's population. And much of that science – in the realm of health, for example – neglects the problems that afflict most of the world's people."

What difference does this disparity in academic output make? Should only universities and research centers be concerned? I do not think so. Consider this interesting correlation. The United States' contribution to the world's annual economic output is between 30 and 40 percent, comparable to its share of scientific output on a global scale. Europe's annual economic output registers a similar percentage and, like the United States, its economic output tracks its contribution to its output of scientific and technological contributions. It is unlikely that this correlation is coincidental.

If we are aware of these trends and understand the problems that stand in the way of progress, why do we have such difficulties building scientific capacity in the developing world and putting science to work to improve its economic well-being? First, the developing world must get its house in order. A renaissance in thinking is needed – we need to pay more attention to education and we should invest more in science and technology. The objective is to provide a new work force equipped with 21st century tools of education and skills and with a belief in ethics and team work. We also need to lower the bureaucratic political barriers that stand in the way of success and to rule by laws that allow for the freedom of thought. Women must participate as full partners in our pursuit of knowledge.

The developing world possesses very capable scientists and continues to contribute outstanding scientists unwittingly to the developed world as part of the brain-drain phenomenon. At Caltech, for instance, my own research group is more than 50% Asian. But major reforms of the system are badly needed. Clearly, this may not be possible on a grand scale in a short time, but the foundation must be established properly and in a timely manner. Empty slogans, or waiting for the developed world to solve the problems, or blaming people of the developed world with conspiracy theories will not provide the means. Yes, international politics play a role, but people's will is a stronger force, provided the force is coherent and not dispersed by internal politics.

The developed world must also carry important responsibilities for its share in partnership, in building the scientific and human capacity in the developing world. First and foremost it must reform its international aid programs, investing less money on military hardware and instruction and more on scientific training and partnerships. Some of the money spent on fighter planes, not to mention the recent war, could fund research programs all over developing countries, helping in what must be the ultimate goal – global education and prosperity through developments. Politics, moreover, should be drained from international aid programs to ensure money is available for productive initiatives that could help boost science and technology in the developing world.

What will rich countries receive in return for the help they give the have-nots? First, there is the moral dimension. The psychological value derived from being a generous global neighbor should not be underestimated. Even on a personal level, most of us do try to help each other, and all major religions encourage and legitimize helping the needy. It is also important to realize that the prosperity of the developed world is in part due to natural and human resources from the developing world, and their markets.

Second, the developed world should acknowledge the importance of reciprocation over time. Islamic civilization gave a great deal to Europe, especially during the Dark Ages. The Arab and Islamic civilizations were major contributors to the European Renaissance. The Islamic civilization rose to become the foremost economic power in the world, at which time it also reached the highest level in the sciences. Today the Muslim world needs help and there is nothing wrong with the United States, Europe, Japan, and other developed nations lending a hand as a modest gesture to the changing fortunes of history.

Third, there is a practical, self-centered consideration based on the time-tested importance of having an adequate insurance policy. In the United States, I pay a great deal for insurance to protect my family against the high cost of medical care, to protect our house against fire and theft, and to protect our cars against accidents. Similarly, the developed world needs to invest in an insurance policy to help it live in a safer and more secure world, but it better be a genuine and good policy!

The choice for the haves is clear. They have to be involved. The choice for the have-nots is also clear. They have to first get their house in order, and build the confidence for a transition to a developed-world status. The transition is possible. In a meeting with Prime Minister Mahathir Mohamad on a recent visit to Malaysia, I learned of the critical role of the new education system implemented during the nation's rapid transition from a labor-intensive economy dependent on cheap labor to a knowledge-based economy poised on the doorstep of the developed world. It is a transition that has been fueled by the belief in building the proper base for modern technology.

The 21st Century – Future Frontiers

Technology in the 21st century is knowledge-based, and unskilled cheap labor, which may have worked for developing countries in the past, will not work in this century. How can the developing world embrace economy-transforming technologies like microcomputing, genetic engineering and biotechnology, information technologies, and femto and nano-technologies without a strong foundation in science? Does the developing world always have to wait decades before participating in global science and technology? Can nations become a part of the modern world without losing their cultural and religious identities? The new century promises unlimited opportunities in science and technology and I believe that the developing world can and should be a partner in or a part of that development. I would like to mention today the new frontiers encompassing three scales in our universe and which I recently outlined:

Our matter – the scale of the very small. We are on our way to being able to manipulate matter at its smallest, most fundamental limits both in time, on the femtosecond scale, and in length, on the nanoscale. Just think about these new scales of time and space in the world of the very small. If your heart beats once a second, now we can see the beats of atoms in a femtosecond, in a millionth of a billionth of a second – a femtosecond is to a minute as a minute is to the age of the universe. Similarly, we can study matter on the nanometer scale and resolve the atoms in their structures – the size of the atom to the size of the earth is like the size of the earth to the whole universe. The opportunity is huge for acquiring new knowledge and for creating new forms of "our matter". The manipulation of matter to produce new sources of energy (photovoltaic/photosynthetic, etc.) should become a major undertaking. The interface of matter's micro- and nanonetworks, designed to produce artificial intelligence, to our life organs, such as the brain, will be another frontier that could alter the boundaries and meaning of species.

Our universe – the scale of the very big. In this century, we may have colonies on the moon, and we may have our second homes on other planets and maybe even in other solar systems. Just think of the scales of the world of the very big. Our universe is about 12 billion years old, and at the speed of light (300,000 km/s), our universe's limit of distance is 100 billion trillion kilometers – certainly enough space for the six billion people on earth today, even multiplying by ten or by one million in the future! The opportunities involving outer space and information technology are unlimited. On our planet, any information one needs will be provided through the so-called "virtual walls" and education and intelligence in all societies will have to be redefined.

Our life – the scale in between. In the first year of this century, the sequencing of the human genome was completed. We now have the genetic map that describes every human on planet Earth. Just think, three billion letters have been deciphered and read into our book of life. The history of biology has changed from the classification of living organisms (Darwin's theory), to the world of cells (Leeuwenhoek–Hooke microscope), to now the molecular world (Watson and

Crick's DNA) with revolutionary ideas in genetic engineering and biotechnology. Soon we might see a nanoscale motor entering the cell to do work. Medicine and human health will certainly enter a new age.

It seems to me that indeed opportunities are unlimited and with education, skill, and the brain power of people around the globe, we should be able to benefit and share in the wealth of new discoveries and technological advances. But, we must first learn how to live together on one globe.

Confluence of Civilizations

The developed and developing nations, aside from their economic and political ties, have to participate in a dialogue among civilizations and a dialogue among cultures. Some intellectuals have introduced concepts such as the "clash of civilizations," as termed by Samuel Huntington, and the "end of history," as expressed by Francis Fukuyama. Both authors argue their cases with conviction, nonetheless, these ideas are controversial and debatable.

As a scientist, I find no "fundamental physics" in these concepts. In other words, it is not a fundamental principle of civilizations that they be in a state of clash with each other. Neither is it a fundamental principle to end history with one system over all other ideologies. I argue that the current world disorder results in part from ignorance about civilizations – unawareness or selective memory of the past and lack of perspective for the future – and in part from the economic misery and political injustices (domestic and/or international) experienced by the have-nots.

According to the dictionary, civilization means an *advanced state* of human society in which a high level of culture, science, industry, and government has been reached. Individually, we are civilized when we reach the advanced state of being able to communicate with and respect those of different customs, cultures, and religions. Collectively, we speak of globalization as a means for bringing about prosperity in the world, yet globalization cannot be a practical concept if there are clashes of civilizations. Historically, there are many examples of civilizations that have coexisted without significant clashes.

I have written about these issues and it is perhaps useful to distill the main points here. The central argument in the thesis of the clash of civilization is that in this post – Cold War era, the most important distinctions among peoples are not ideological, political, or economic – they are cultural. Hence people define themselves in terms of ancestry, religion, language, history, values, customs, and institutions. According to this thesis the world becomes divided into eight major civilizations: Western, Orthodox, Chinese, Japanese, Muslim, Hindu, Latin American, and African.

I have several difficulties with this analysis, and perhaps the following questions and commentary may clarify my position. First, *what is the basis for these divisions of civilizations?* People belong to different cultures, nations have experienced (and continue to experience) different cultures, and nations on the same continent may be influenced by different civilizations. In my own case, from birth to the present time, I can identify myself as Egyptian, Arab, Muslim, African, Asian, Middle Eastern, Mediterranean, and American. Looking closely at just one of these civilizations, I note that the Egyptian people belong to a dynamic civilization with a multicultural heritage: Pharaonic, Coptic, Arabic, Islamic, not to mention the Persian, Hellenistic, Roman, and Ottoman influences.

Second, *is it fundamental that differences in cultures necessarily produce clashes?* In this thesis, it is contended that if the United States loses its European heritage (English language, Christian religion, and Protestant ethics), its future will be endangered. I reach the opposite conclusion. From a personal point of view, I did not speak English when I came to the United States, I am not a Christian, and I was not taught Protestant ethics. Yet I integrated into my new, American culture while preserving my native culture(s) and I believe that the confluence of my "Eastern" and "Western" cultures is without a clash. From a broader perspective, America's strength has traditionally risen from its "melting pot" culture; the country has been enriched and continues to be enriched by multiethnicity and the different cultures of its inhabitants.

Turning to international relations, cultures and civilizations can be at their peak of achievement and yet coexist in harmony and even complement each other. The United States, Japan, and European nations are examples of this beneficial coexistence created by building economic and cultural bridges. Another example comes from a country that many would have doubted had the potential for creating ethnic and religious harmony: Malaysia with its inhomogeneous population of Malays (53%), Chinese (26%), and Indians (8%) with different religions – Muslim (60%), Buddhist (19%), Christian (9%) and Hindu (6%). Neither religion nor culture seems to hinder its progress, and certainly Malaysia is an economic success story, demonstrating the civility of living harmoniously together with a variety of cultures.

Finally, *what about the dynamics of cultures?* Cultures are not static; they all change with time, and the degree of change is governed largely by forces of politics, economics, and religions. Let us consider my home country. Egypt's civilization was developed very early in human history and it dominated the world for millennia, but lately the nation has become a developing one. This does not mean that Egypt has lost its civilization, but it does mean that, like others, it has changed with time, due to many internal and external forces – the current state is not due to a genetic factor or fundamental cultural values. Cultural changes may impede progress within a nation, but not necessarily through clashes with others, or permanently.

What we have to consider seriously are the political and economic interactions within a culture and between the various cultures of the world. The people of North and South Korea are of similar culture, but the notable disparity in progress between the two countries is due to economic and political factors; the same can be said of East and West Germany before reunification. It is easy to divide the world into "us" and "them", and slogans such as the clash of civilizations or the conflict of religions certainly make it very difficult to unite the nations of the world – we need dialogues, not conflicts or clashes!

Epilogue

I would like to close with a message. The world at the beginning of the 21st century is divided, not only politically but also between hope and hopelessness. On one hand, progress can be seen in the human life – life expectancy has increased by ten years over the past three decades, infant mortality has fallen by 40%, and adult illiteracy has been reduced by half. On the other hand, every day 30,000 children die of preventable diseases, some 60 countries grow poorer, the spread of HIV/AIDS has become the most deadly epidemic in human history according to the UN, and the crisis of global water is becoming real as we witness in Iraq today and most probably in future conflicts on the whole planet. It is not naive to think of a better world and to achieve that goal through courage, justice, and liberty.

Judging from history, the shape of the future is made by leaders who have the capacity to turn it into an epoch of hope for peace and prosperity or into one of divisiveness and disorder. It is unlikely that those who do not know about history will make history. Leaders of the world should use the benefits of knowledge to shape a hopeful future for our children and grandchildren, for posterity. This can only be achieved by understanding the need for justice in the world and by promoting dialogues and cooperation among countries and peoples of the world, the world community. That is why it is vital to maintain a financially strong and independent UN. Despite the complexity in world affairs, undermining the UN as a viable institution for world education and peace will be a tragedy of enormous consequences. Even the superpower, the United States, whose population is only 5% of that of the total on planet Earth, cannot be the world's judge, jury, and executioner.

To shape our future in the age of globalization we need to develop a new perspective – one encompassing the economic, political, religious, and cultural dimensions of world affairs. Miss Kalpana Chawla, an Indian-born naturalized American who lost her life in the space shuttle Columbia disaster on February 1st of this year (2003) said, "When you are in space and look at the stars and the galaxy, you feel that you are not just from any particular piece of land, but from the solar system." She was viewing the world from the heavens and she had a universal perspective. A true statesman or stateswoman will see our world with a universal perspective that is unifying for humanity. And then wars may become wars on global poverty, disease, and despair, and for a sustainable world future.

References

[1] Øivind Stenersen, Ivar Libaek, Asle Sveen, *The Nobel Peace Prize*, Cappelen Forlag AS, Oslo (2001).

[2] The Economist, February 8, 2003 (for Ms. Kalpana Chawla); The World in 2003 (for Kofi Annan's MDGS); April 5, 2003 (for Malaysia's Report).

[3] Kofi Annan, Science, Volume 299, page 1485 (2003).

[4] Ahmed Zewail, Science for the Have-Nots, Nature, (London 2001), Vol. 410, p. 741.

[5] Ahmed Zewail, Dialogue of Civilizations, SSQ2/Journal, Routledge Press, Paris, France (2002); Address at UNESCO, April 20, 2002.

[6] Ahmed Zewail, Voyage through time – Walks of Life to the Nobel Prize, American University in Cairo Press (2002); Reprinted in 12 languages and editions.

Ahmed Zewail

Ahmed Zewail, was born in Egypt and studied at the University of Alexandria and University of Pennsylvania. He is presently the Linus Pauling Chair Professor of Chemistry and Professor of Physics, and the Director of the NSF Laboratory for Molecular Sciences (LMS) at Caltech, the California Institute of Technology, Pasadena, USA. Zewail has received numerous Prizes and Awards including the 1999 Nobel Prize in Chemistry for his pioneering developments in the field of femtoscience. Professor Zewail is also renowned for his tireless efforts to help the population of the have-nots.

Einstein and War Resistance

Eva Isaksson

When the First World War broke out, Einstein had just settled down in Berlin at the Kaiser Wilhelm Institute. He was already a well-known physicist despite his young age. Born in Germany, Einstein had studied in Switzerland and preferred its freer intellectual atmosphere to the extent that he had taken Swiss citizenship at the age of 16. There is no evidence of his interest in social affairs before the war. He was a loner who avoided authority, and valued intellectual freedom quite highly.

There had been a strong rise in international cooperation in science since the late 19th century. International scientific meetings flourished, and there was for some time an atmosphere of fruitful international exchange of scientific ideas. The creation of Nobel prizes reflected an international appreciation of scientific achievements. At the same time, the peace movement was building its international structures. The International Peace Bureau was formed in 1891 to coordinate international peace gatherings. Its approach was grounded in bourgeois liberalism, seeking the peaceful arbitration of international conflicts. The well-known IPB activist and vice-president Bertha Suttner influenced Alfred Nobel to start a Nobel peace prize, thus making a link between scientific work and peace activism. However, the scientists' awareness of the reasons for war, and their own role in it, were not well developed in 1914. When the war started, its decisive difference from all the previous wars was not immediately evident. It was the first total war, in which mass weapons created by technological innovations killed civilians as well as armies.

The majority of professors at the Kaiser Wilhelm Institute lacked the cosmopolitan perspective that Einstein had. United by their nationalist sentiments, they enthusiastically supported the German war effort, either at the front or as military research experts. A sign

Einstein – Peace Now! Reiner Braun and David Krieger (Eds.)
Copyright © 2005 WILEY-VCH Verlag GmbH & Co. KGaA, Weinheim
ISBN 3-527-40604-2

of these spirits was the "Manifesto to the civilized world", published in the autumn 1914, and also called "Manifesto 93", as it was signed by 93 scientists. Einstein reacted adversely to this strongly nationalist manifesto. Together with a well-known pacifist, professor of physiology Georg Friedrich Nicolai, he drew up the "Manifesto to the Europeans", which states: „The struggle raging today can scarcely yield a "victor"; all nations that participate in it will, in all likelihood, pay an exceedingly high price."[1] This manifesto was signed by only two others and it was never published, but in any case it was the first anti-war manifesto signed by Einstein.

The new anti-war movement began to regard social conditions as a cause of war. Its supporters were working in difficult circumstances and separated from each other. Open war resistance often led to imprisonment. In 1914 Einstein joined a small antiwar group, the Bund Neues Vaterland. Its membership was prominent – lawyers, aristocrats, even bankers. The group was forbidden in 1916 but continued its activities underground, to be revitalized after the war. Its main goal was to create a supranational organization that would make future wars impossible. This was to be a solution to the problem of war and peace, which Einstein would support for the rest of this life.

In September 1915, Einstein met Romain Rolland in Switzerland. Rolland, a well-known peace activist, describes this meeting in his diary:

> (...) Einstein is incredibly outspoken in his opinion about Germany, where he lives and which is his second fatherland (or his first). No other German acts and speaks with a similar degree of freedom. Another man might have suffered from a sense of isolation during that terrible last year, but not he. He laughs. He has found it possible, during the war, to write his most important scientific work. I ask him whether he voices his ideas to his German friends and whether he discusses them with them. He says no. He limits himself to putting questions to them, in the Socratic manner, in order to challenge their complacency. People don't like that very much, he adds. 'Greedy' seems to Einstein the word that best characterizes the Germans. Their power drive, their admiration of, and belief in, force, their firm determination to conquer and annex territories are everywhere apparent. ... The socialists are the one relatively independent element. ... The Bund Neues Vaterland makes rather slow progress and does not enjoy wide support. ... Einstein does not expect that Germany will be reformed under its own power. ... He hopes for an allied victory, which would destroy the power of Prussia and its dynasty. [2]

The war ended just as Einstein had hoped – in an allied victory. The spirits among those who believed that the peace would now solve the burning questions facing mankind were quite high. After the war, Einstein stood sympathetic towards the Social Democratic Party, which he never joined, however. He did not like its faltering attitude towards the war, and the failure of its politics during the Weimar republic alienated Einstein from party politics. He sympathized with socialists, but had no formal ties to them. After the bending of light during the eclipse of 1919 had been detected, Einstein had become world famous, and his opinions gained weight. He did not hesitate to use this to further the cause of peace. His name began to appear in appeals for a variety of causes.

The Weimar republic proved a disappointment for those hoping for peace and democracy. Einstein was a Jew and a socialist, and as such a good target for reactionary forces. There were numerous attempts to discredit both his scientific achievements and his political opinions. His scientific colleagues were also critical of his views. When Einstein refused to accept some of the fundamental assumptions of the new quantum mechanics, some scientists remarked that he was putting so much effort into his activism that he was not able to follow the progress of modern science.

The League of Nations was founded in 1920, and at first seemed to represent the goals that Einstein had expected from a world government. The League was, however, formed by the victors, and the former central powers, such as Germany, were not treated equally. The International Peace Bureau supported the League, however, demanding its reform. In the 1920s, the peace movement was quite strong, and also split into more and less radical groups. At that time, Einstein sympathized with the radical peace activists, only to grow more conservative in his views a decade later.

The League of Nations called in 1922 some well-known scientists and scholars, Einstein among them, to form the Committee on Intellectual Cooperation. Einstein was full of doubts already from the start. Would he be representing Germany in a body run by those representing the victorious side? Was the League of Nations really doing work that was worth supporting? Einstein began to express increasing doubts about these matters. The World Disarmament Conference, which was in preparation from 1926, but that was held as late as 1932, was the last straw for Einstein. He felt that the conference was still only considering possibilities for limiting armaments when Germany was already leaving those talks on the eve of Hitler's rise to power. Einstein withdrew from the committee in 1932 with a public statement criticizing the League of Nations, stating: "One does not make wars less likely to occur by formulating rules to warfare."

Einstein was still a radical pacifist in the 1920s: "My pacifism is an instinctive feeling that possesses me; the thought of murdering another human being is abhorrent to me. My attitude is not the result of an intellectual theory but is caused by a deep antipathy to every kind of cruelty and hatred. ..."[3] From this starting point, Einstein was ready to condemn every form of military service no matter what he might think of the causes of any particular war. He did not, however, give any visible public support to militant pacifist movements before 1928.

In 1930, touring the US, Einstein made a well-known speech that represents the culmination of his pacifist ideas. He made this so-called '2 per cent' speech on December 14 1930 in a meeting arranged by the New History Society. In this speech, made without notes, he argued that even if only two per cent of those assigned to perform military service should announce their refusal to fight, as well as urge means other than war of settling international disputes, governments would be powerless, they would not dare send such a large number of people to jail. In Einstein's opinion, conscious objectors should be permitted to do "some strenuous or even dangerous work, in the interest of their own country and mankind as a whole."[4]

This speech was met with enthusiasm by pacifists, and Einstein quickly became an international hero in the eyes of the peace movement. There were however also critical comments, such as the remark made by Romain Rolland:

> Einstein seems to overlook the fact that the technique of war has changed since 1914, and is still changing. The tendency has been to employ small armies of technicians who know how to run air squadrons armed with gas and bacteriological torpedoes and other weapons of mass destruction. In such circumstances it becomes a matter of complete indifference to governments whether two or ten per cent of the population refuses military service. Governments would not even need to throw war resisters into jail. Soldiers and non-combatants alike would be subjected to the deadly rain (...). [5]

Another disputed point was which kind of work could be considered non-military. In the long run, any work could be considered to benefit the war effort. In 1929, Einstein signed a letter drafted by the War Resisters International to free the Finnish conscientious objector Arndt Pekurinen, who had been jailed for his refusal to accept any form of military service. When the Finnish Minister of Defence replied that the conscription law allowed the employment of conscientious objectors for non-military work, Einstein sent a congratulatory letter, which received wide publicity in Finland. Upon receiving clarifying information from Finnish peace activists, Einstein again wrote a letter to the Finnish minister: „Morality and fairness make it necessary that those who object to military service be employed only in work that has no close connection with military purposes."[6] The publicity that Einstein's letters added to this case may have speeded up the new Finnish conscription law of 1931.

In 1932, before the breach with the League of Nations, the Committee on Intellectual Cooperation had invited Einstein to send an open letter to Sigmund Freud. Their exchange was published as a booklet "Warum War?" (Why war?) in 1933. Only 2000 copies were printed, and after Hitler's rise into power the distribution of this booklet was prohibited.

In his letter Einstein asked whether there is any way of delivering mankind from the threat of war. How is it possible that a small clique can bend the will of a majority who stand to lose and suffer by a state of war, to the service of their ambitions? How do they succeed so well in rousing men to such wild enthusiasm, even to sacrifice their lives? Einstein asks: "Is it possible to control man's mental evolution so as to make him proof against the psychosis of hate and destructiveness?"[7]. Freud's reply was quite pessimistic. According to him, conflicts between men are resolved with violence, and cultural evolution has just tried to suppress brute force by transferring power to a larger combination founded on the community of sentiments linking up its members. A central controlling body, like the League of Nations, would only succeed in a peacekeeping task if it has force at its disposal.

At the time of Hitler's rise to power in winter 1933 Einstein was in the US and decided not to return to Germany. In March he stated:

As long as I have any choice in the matter, I shall live only in a country where civil liberty, tolerance and equality of all citizens before the law prevail. Civil liberty implies freedom to express one's political convictions, in speech and in writing; tolerance implies respect for the convictions of others whatever they may be. These conditions do not exist in Germany at the present time. [8]

The events in Germany affected Einstein's views on antiwar measures. He decided that Germany's threat to world peace was so great that passive resistance was not enough. Einstein spent his last months in Europe in Belgium. He was approached by Alfred Nahon, a Belgian lawyer asking him to support two conscientious objectors. Einstein refused, and also published his reasons for doing so in an open letter, which caused a stir among pacifists. Einstein writes: "Were I a Belgian, I should not, in the present circumstances, refuse military service; rather, I should enter such service cheerfully in the belief that I would thereby be helping to save European civilization."[9] Already prior to this Einstein's support to World Government had caused resentment. Pacifists could not accept his reasoning that the World Government should have military resources. "Other times, other means, although the final goal remains unchanged," says Einstein in his 1934 article "Re-examination of pacifism".

Pacifists had reason to feel that Einstein had betrayed them, as his name had helped to attract credibility to the peace movement. Thousands of conscientious objectors had gained strength from this vocal support. Reactions could be quite bitter. Romain Rolland writes in his diary:

Such weakness of spirit is indeed unimaginable in a great scientist, who should weigh and express his statements carefully before putting them in circulation. It is even more incredible coming from the author of the Theory of Relativity. Had it never occurred to him that circumstances might develop, circumstances such as those that prevail today, which would make it dangerous to practice conscientious objection which he espoused? It is a joke, a kind of intellectual game, to advocate the idea at a time when no risks are involved; on the other band, one has assumed a particularly serious responsibility for having indoctrinated blind and confident youth without sufficient consideration of all implications. It is quite clear to me that Einstein, a genius in his scientific field, is weak, indecisive and inconsistent outside it. I have sensed this more than once. (...) One can imagine the homicidal fury of the Hitlerites when they learned that a German had sounded the call to arms to other nations against Germany. Nothing could have been more fatal to the cause of the Jews in Germany. Einstein did not anticipate this. I am afraid that he may now find it quite difficult to justify himself. His constant about-faces, hesitations and contradictions are worse than the inexorable tenacity of a declared enemy (...) [10]

Faced with such accusations, Einstein did not defend himself, but raised instead a desperate question: „Can it be that the world does not see that Hitler is dragging us into war?"[11]

In 1934 Einstein left for the US and never returned to Europe. He settled down in Princeton at the Advanced Study Institute. Einstein was not, however, willing to fully concentrate on scientific work, and continued his political activities.

World War II was an overwhelming challenge to peace activists. Einstein met this challenge with the perhaps biggest miscalculation in his career, the letter to President Roosevelt that Einstein composed together with Leo Szilard, and that was dated August 2, 1939. His apprehension towards Hitler's Germany was so great that Einstein felt he must warn the US government about the possibility about uranium deposits in Belgian Congo falling in German hands. At that time, Einstein and many others felt they had reason to fear that German was developing an atom bomb. Roosevelt took these warnings seriously, and the Briggs committee was started. Not satisfied with the slowness with which this committee seemed to work, Einstein wrote another letter to the President in March 1940, repeating his warnings.

After these two letters, Einstein had no further connection with the atomic bomb project. It was probable that he was well aware of the progress of its development, however. Szilard contacted Einstein once again in early 1945, expressing his worries about the use of bomb, and how it would affect the post-war situation. Einstein wrote his third letter to the US President, dated March 25, 1945.

The terms of secrecy under which Dr. Szilard is working at present do not permit him to give me information about his work; however, I understand that he now is greatly concerned about the lack of adequate contact between scientists who are doing this work and those members of your Cabinet who are responsible for formulating policy. [12]

Roosevelt died before Leo Szilard presented this problem to him. His successor, President Truman, was not affected by Einstein's letter in any way. The US made a successful atom bomb test in July, and by the time the bomb was to be deployed against Japan there was already a considerable number of scientists opposed to its use. An appeal initiated by Szilard and signed by sixty scientists, condemning the military use of the atomic bomb, was unsuccessful, and the US dropped the bomb on Hiroshima on August 6 and another bomb on Nagasaki on August 9.

> My participation in the production of the atomic bomb consisted of one single act," wrote Einstein later. "I signed a letter to President Roosevelt, in which I emphasized the necessity of conducting large-scale experimentation with regard to the feasibility of producing an atom bomb. I was well aware of the dreadful danger which would threaten mankind were the experiments to prove successful. Yet I felt impelled to take the step because it seemed probable that the Germans might be working on the same problem with every prospect of success. I saw no alternative but to act as I did, although I have always been a convinced pacifist. [13]

It is certain that the atomic bomb would have been developed without Einstein's intervention. It remains an open question whether the bomb would have been finished before the end of the second World War, and how the possible delay would have affected the post/war world situation.

"I do not consider myself the father of the release of atomic energy," said Einstein in his first public talk about the atomic bomb, given to Raymond Swing, and published in Atlantic Monthly in November 1945, under the title "Atomic War or Peace". "My part in it was quite indirect. I did not, in fact, foresee that it would be released in my time. I only believed that it was theoretically possible. It became practical through the accidental discovery of chain reaction, and this was not something I could have predicted."[14]

This was the beginning of a long series of writings that Einstein composed against the bomb and against the arms race. His own solution to the problem of war was now world government, which he saw as an organization like the United Nations and that would have both political and military power. This new world government was urgently needed, so that total destruction could still be avoided. Einstein joined organizations like the Emergency Committee of Atomic Scientists. His name attracted funds and publicity, although his views were often more radical than most of his fellow scientists. He was particularly vocal about the necessity of involving the Soviet

Union in the international community. To begin with, it should receive observer status in the new world government. Einstein thought that the Soviet Union was feeling threatened by the growing military power in the west, and that it should become involved in the peace process on a more equal basis. Einstein's views received heated criticism from Soviet scientists, who composed an open letter on "Dr. Einstein's mistaken notions", in which they said: "By the irony of fate, Einstein has virtually become a supporter of the schemes and ambitions of the bitterest foes of peace and international co-operation." This exchange reflects the rift between peace activism in the west and in the east. Many peace activists in the west were also split about the issue of world government, among them the International Peace Bureau, which consequently underwent a quiet phase in these difficult post-war times. In the east, peace activism was channelled to the World Peace Council, which was partial to the Soviet Union and funded by it.

The Emergency Committee of Atomic Scientists was still supportive of world government in April 1948 when it published a declaration supporting it, but had given up the urgency of achieving it in the near future. Einstein never gave up the idea of a world government. When a pacifist asked him in 1952 about his real stand regarding pure pacifism, Einstein replied:

> I am indeed a pacifist, but not a pacifist at any price. My views are virtually identical with those of Gandhi. But I would, individually and collectively, resist violently any attempt to kill me or to take away from me, or my people, the basic means of subsistence. …I was, therefore, of the conviction that it was justified and necessary to fight Hitler. For his was such an extreme attempt to destroy people. … Furthermore, I am of the conviction that realization of the goal of pacifism is possible only through supranational organization. To stand unconditionally for this cause is, in my opinion, the criterion of true pacifism. [15]

In the divided post-war atmosphere, Einstein became a solitary thinker. During the Cold War years, he was often under attack. Reactionary circles resented his critical views about the US foreign and domestic politics and his negative approach towards anti-communism. Einstein took a strong stand against the Committee on Un-American Activities. In a letter published in New York Times in June 1953 he stated: "Every intellectual who is called before one of the committees ought to refuse to testify, i.e. he must be prepared for jail and economic ruin, in short, for the sacrifice of his personal welfare in the interest of the cultural welfare of his country."[16] Einstein's statement gave many reactionary politicians a reason to accuse him of incitement to civil disobedience. Einstein said in 1964: "The Communist menace is being used here by reactionary politicians as a pretext to mask their attack on civil rights."[17]

In 1950, the United States made a decision to start the development of the hydrogen bomb. When Einstein was asked for a statement, he said: „I do not believe your proposal that the United States refrain from experimenting with the production of hydrogen bombs touches the core of the problem. The fact of the matter is that the people who possess the real power in this country have no intention of ending the Cold War."[18] Einstein would most likely have expressed similar views about the world situation of the early 2000s, and have called for a stronger United Nations to further world peace in a world where the United States has shown no willingness to give up its dominant role in world politics in favor of a more balanced world order. In a commencement address in Swarthmore College in 1938, Einstein gave a clear expression of his views about a socially moral individual:

> Were he to receive from his fellowmen a much greater return in goods and services than most other men ever receive? Were his country, because it feels itself for the time being militarily secure, to stand aloof from the aspiration to create a super-national system of security and justice? Could he look on passively, or perhaps even with indifference, when elsewhere in the world innocent people are being brutally persecuted, deprived of their rights or even massacred? To ask these questions is to answer them! [19]

The Emergency Committee of Atomic Scientists became inactive in the late 1940s. The world situation had not proved benign to world government. During the Cold War years, there seemed to be practically no means to avoid nuclear catastrophe. In February 1955 Bertrand Russell sent Einstein a letter that opened new perspectives in anti-war activism. "Do you think it would be possible to get, say, six men of the very highest scientific repute, headed by yourself, to make a very solemn statement about the imperative necessity of avoiding war?" Russell stressed: "Everything must be said from the point of view of mankind, not of this or that group."[20]

Einstein replied quickly and full of enthusiasm, making suggestions about possible signatories. After a short and rapid exchange of letters, a final version was drafted which Russell sent Einstein on April 5, 1955. Einstein wrote back on April 11, approving the list of other signatories and expressing his consent to sign the appeal. These were the last documents that Einstein signed before his death. Russell heard about Einstein's passing on his way from Rome to Paris. Upon his arrival, he received Einstein's last letter and the declaration signed by him.

The declaration, which was published later that year, was signed by eleven scientists, of which nine were Nobel laureates. "We are speaking on this occasion, not as members of this or that nation, continent or creed, but as human beings, members of the species man, whose continued existence is in doubt," the declaration states. It had to be understood that the nature of war had changed, so that no one would gain but instead, all participants would face destruction.

In view of the fact that in any future world war, nuclear weapons will certainly be employed, and that such weapons threaten the continued existence of mankind, we urge governments of the world to realize, and to acknowledge publicly that their purposes cannot be furthered by a world war, and we urge them consequently, to find peaceful means for the settlement of all matters of dispute between them. [21]

The declaration was published right before two major international conferences, and received much attention. It was the first anti-war declaration signed by this number of leading scientists, many of which represented differing political views. The difference with the situation at the outbreak of the World War I was huge. Not only had the military technology taken a leap forwards, but also the anti-war consciousness of scientists had come a long way forward.

An immediate continuation of the Russell-Einstein manifesto was the first Pugwash conference of scientists that met in Canada in 1957, and that started the series of Pugwash conferences. It has involved a large number of scientists from many countries and disciplines. What makes the Pugwash movement special is that it was started by the scientists out of their own initiative. Like Einstein, they have understood that peace will benefit all. Scientists have a special role in the peace movement because they are able to predict not on-

ly the consequences of scientific inventions, but also possibilities of
their application.

References

[1] Otto Nathan and Heinz Norden, (Eds.), *Einstein on Peace*, London, Methuen, 1963, p. 5.
[2] Ibid., p. 14–16.
[3] Ibid., p. 98.
[4] Ibid., p. 117.
[5] Ibid., p. 118.
[6] Ibid., p. 128.
[7] Ibid., p. 189–190.
[8] Ibid., p. 211.
[9] Ibid., p. 229.
[10] Ibid., p. 232–233.
[11] Ibid., p. 235.
[12] Ibid., p. 305.
[13] Ibid., p. 584.
[14] Ibid., p. 350.
[15] Ibid., p. 564.
[16] Ibid., p. 547.
[17] Ibid., p. 602.
[18] Ibid., p. 519.
[19] Reproduced in: Swarthmore College Bulletin, December 2002, Otto Nathan and Heinz Norden, (Eds.), l.c., http://www.swarthmore.edu/bulletin/dec02/em_einstein.html.
[20] Ibid., p. 625.
[21] Ibid., p. 628.

Eva Isaksson, IPB

Eva Isaksson, physicist, works at the University of Helsinki. The International Peace Bureau is the worlds' oldest and most widespread international peace-federation, bringing together peaple working for peace in many different sectors. The IPB was awarded the Nobel Peace Prize in 1910.

The Military Mentality

John Stachel

In its Summer 1947 issue, The American Scholar published an article by Albert Einstein entitled "The Military Mentality," a response to an article by Louis Redenour in the Spring issue of that magazine entitled "Military Support of American Science, A Danger?" essentially denying that such a danger existed.[1]

In his reply, Einstein soon passes beyond the question raised by Ridenour to what he sees as the issue at the root of all such questions: the military mentality:

> It is characteristic of the military mentality that non-human factors (atom bombs, strategic bases, weapons of all sorts, the possession of raw materials, etc.) are held essential, while the human being, his desires and thoughts – in short, the psychological factors – are considered as unimportant and secondary ... The individual is degraded to a mere instrument; he becomes 'human materiel.' The normal ends of human aspiration vanish with such a viewpoint. Instead, the military mentality raises 'naked power' as a goal in itself – one of the strangest delusions to which men can succumb. [1]

He points out that:

> The Germans, misled by Bismark's successes in particular, underwent just such a transformation of their mentality – in consequence of which they were entirely ruined in less than a hundred years.[1] [2]

Einstein knew whereof he spoke. He had witnessed the culmination of German militarism during the First World War (1914–1918) from Berlin, the heart of the German Reich; and from this same vantage point he lived through the first stages of German ruination: military defeat and aborted revolution in 1918–1919 and subsequent

---

1) From the time of Bismark's triumph in the Franco-Prussian War of 1870–1871 to the end of World War II in 1945, only three quarters of a century had elapsed.

Einstein – Peace Now! Reiner Braun and David Krieger (Eds.)
Copyright © 2005 WILEY-VCH Verlag GmbH & Co. KGaA, Weinheim
ISBN 3-527-40604-2

rise, decline and fall of the Weimar republic (1919–1932). After the Nazi seizure of power in 1933, he severed all ties with Germany and from Princeton, N.J., watched the rise and fall of the Third Reich (1933–1945) culminating first in the spiritual ruination of the German people under fascism, and then their physical ruination in the last days of World War II. He now expressed the fear that his new homeland was embarking on the same path:

> I must frankly confess that the foreign policy of the United States since the termination of hostilities [in 1945] has reminded me, sometimes irresistibly, of the foreign policy of Germany under Kaiser Wilhelm II, and I know that, independently of me, this analogy has most painfully occurred to others as well.

With remarkable prescience, only two years into the Cold War, he foresaw where this trend was leading the United States:

> Today, the existence of the military mentality is more dangerous than ever because the offensive weapons have become much more powerful than the defensive ones. [This was written after the development and use of the atomic bomb (1945), but before the development and testing of the hydrogen bomb (1951)]. This fact will inevitably produce the kind of thinking that leads to preventive wars. The general insecurity resulting from these developments results in the sacrifice of the citizen's civil rights to the alleged welfare of the state. Political witch-hunting and governmental controls of all sorts (such as control of teaching and research, of the press, and so forth) appear inevitable, and consequently do not encounter that popular resistance that, were it not for the military mentality, might serve to protect the population. A reappraisal of all traditional values gradually takes place and anything that does not clearly serve the utopian goal of militarism is regarded and treated as inferior. [3]

By extrapolating the trends in the United States that he saw in the 1940s in the light of his experience of German militarism, Einstein was able to predict with uncanny accuracy the contemporary situation we face in the United States. When reading his words, who can avoid thinking of the elusive "war on terror" or the all-too-concrete wars on Afghanistan and Iraq, with their mounting list of American war crimes; of the unpatriotic "Patriot Acts," so reminiscent of the Alien and Sedition Laws that blighted the lives and liberties of an earlier generation of Americans and their foreign guests; of the careful management of information – and mismanagement of people – the sedulous spread of misinformation by government agencies charged with facilitating our rights to information, to liberty and true security? And behind this smokescreen of manipulated fear and provoked panic the unremitting assault on the living standards of the working people of this land goes on: the attempt to divide and conquer all by setting organized against unorganized, young against old, native born against foreign-born, gay against straight, gender against gender, pro-choice against pro-life – the list goes on endlessly.

I do not mean to imply that these trends had to triumph to the extent that they already have, or that they will inevitably continue to prevail. What I mean is that, as long as popular resistance to them in the United States remains weak and disorganized, these trends will go on unchecked.

In Germany, these trends unchecked led to war, brutality and fascism. In the United States they have already led to wars presented as "preventive," to the indefinite confinement and torture of men and women proclaimed to be prisoners of this war, whether guilty of any crime or not. Will it lead to fascism? There are certainly tendencies in that direction, and many elements in our history that provide the soil for its growth. But we must not forget that we Americans are the inheritors of an ambiguous legacy. As well as a history of wars of conquest and extermination of the indigenous inhabitants of North America, and of the abduction and enslavement of countless people of color, the United States also has a precious history of struggle – often times successful – for their democratic rights by its working people, its minorities and its women. If we can mobilize sufficient popular forces to revive and reinvigorate these traditions of struggle and apply them to our contemporary problems, the tide can still be turned.

And we in the United States must never forget that resistance abroad to American hegemony is an aid in our struggles at home. We must never let a false nationalism take the place of true national feeling, that can only feel shame when it sees American democratic values and traditions trampled and the worst features of our past flaunted.

Even in Germany, as late as 1932, it was still possible to stem the tide of Nazism through united action of all anti-fascist forces, as called for by Einstein, together with Kaethe Kollowitz and Heinrich Mann.[4] There is one factor in our favor that is often neglected. Just as the popular forces are segmented in many ways – by class, by occupation, by gender, by race, by nationality, etc. – the ruling classes are also segmented and their coalition can be broken up under the impact of popular pressure. The reaction to the drive for war on Iraq already revealed fracture lines between the "old" conservatives gathered around the elder George H. W. Bush and the "neo-conservatives" who flocked under the banners of the younger George W. Bush. Unless and until a popular coalition develops that is strong enough – and tactically wise enough – to put sufficient pressure on the fracture lines within the ruling classes (think of how the Vietnam war was ended), there will be no successful resistance to the drive towards war and fascism – let alone a successful counter attack to impose policies of peace and liberty on our government.

In the coming struggle, we the living need all the help we can get from the great heroes of our past. We must never forget the lesson taught by Frederick Douglass: "Without struggle there is no progress." And we should take heed of the warning by Thomas Jefferson of what faces us if we fail: "I tremble for my country when I reflect that God is just; that His justice cannot sleep forever." Through his writings and the force of his moral example, Albert Einstein stands at our side in this struggle.

References

[1] See Otto Nathan and Heinz Norden, (Eds.), *Einstein on Peace*, New York, Schocken, 1960, pp. 422–424 for more details on the context of this article and an English translation. *Albert Einstein, Ideas and Opinions*, New York, Crown 1954, pp. 132–134, reprints another translation, which we shall cite, occasionally correcting it with the use of the first-cited one.

[2] Otto Nathan and Heinz Norden, l.c., p. 421.

[3] Ibid., p. 422.

[4] See their appeal for unity, Doc. 146, in Christa Kirsten and Hans-Jürgen Treder, (Eds.), *Albert Einstein in Berlin 1913–1933*, Vol. 1, Darstellung und Dokumente, Berlin, Akademie-Verlag 1979, p. 223.

John Stachel

John Stachel is a Physicist and was born in 1928. He directs the Boston University Center for Einstein Studies.

The Role of Civil Society in Disarmament Issues: Realism vs. Idealism?[1]

Jody Williams[2]

I want to thank both the UN Department of Disarmament Affairs and the Government of the People's Republic of China for holding this conference on the critical issue of "A Disarmament Agenda for the 21st Century" at this very difficult point in our history. I also want to thank the co-sponsors for inviting me to be here and for giving me the opportunity to speak.

Like Dr. McCoy, I started thinking and writing about what I might say before arriving in Beijing. However, unlike Dr. McCoy, I did not finish in a timely fashion – but for that, I am pleased as I do not find myself, as he said he did, wishing for a chance to re-submit my comments, as I had the luxury of re-thinking my speech at the end of the day yesterday. I am glad for that, because I threw out much of what I had written before arriving.

As you note from the agenda, I am asked to comment on the role of NGOs – nongovernmental organizations, civil society – in the field of disarmament. I confess to feeling somewhat schizophrenic in trying to speak to that topic here. With schizophrenia, various elements compete in the schizophrenic's world to define reality; some of those elements are quite real, but others are real only in the mind of the schizophrenic. Because of this duality, it is difficult for the schizophrenic to know what to react to and how.

After listening to the first day and a half of discussion, I must make another confession – and that is that my sense of what constitutes reality and what is unreal has been further blurred rather than clarified. Since I am not the only speaker here to have used this anal-

1) For A Joint Conference: "A Disarmament Agenda for the 21st Century" cosponsored by The United Nations Department for Disarmament Affairs and the People's Republic of China 2–4 April 2002.

2) Nobel laureate for Peace Campaign Ambassador, International Campaign to Ban Landmines

Einstein – Peace Now! Reiner Braun and David Krieger (Eds.)
Copyright © 2005 WILEY-VCH Verlag GmbH & Co. KGaA, Weinheim
ISBN 3-527-40604-2

ogy, I trust that some might have a sense of what I mean – for the others, I ask your indulgence as I try to explain.

It is obvious, I guess, why I was invited to speak on this subject. I have had the privilege of being involved in a global movement to eliminate antipersonnel landmines that has seen much success. Seemingly out of nowhere, the ban movement was able to build awareness and capture the public conscience in such a way that resulted in government action to deal with landmine proliferation in a timely fashion.

When the Norwegian Nobel Committee chose to recognize the work of the International Campaign to Ban Landmines (ICBL) with the awarding of the Nobel Peace Prize in 1997, it noted that the Campaign had, with the Mine Ban Treaty of the Ottawa Process, made feasible reality of a utopian dream. Further, the Committee noted that the campaign had been able to "express and mediate a broad range of popular commitment in an unprecedented way. With the governments of several small and medium-sized countries taking the issue up ... this work has grown into a convincing example of an effective policy for peace." It concluded, "As a model for similar processes in the future, it could prove to be of decisive importance to the international effort for disarmament and peace."

The ban movement and the ICBL in particular flourished in what now almost seems like a dream moment in the period following the end of the Cold War when anything seemed possible. The momentum of the movement to eliminate landmines grew seemingly effortlessly – in a world no longer bi-polar where any number of alliances seemed possible and where we would call the partnership of civil society and governments seeking to rapidly redress global problems "a new superpower."

Given this experience, my perception and my concrete reality affirm that there is a role for civil society to play in defining a disarmament agenda. Given this experience, my perception and my concrete reality affirm that "realism" does not preclude having a vision of a better world and turning that vision into reality. My "schizophrenic doubts" are quelled for a moment and I can speak about the role of NGOs in disarmament issues.

It was also heartening to hear numerous speakers here recognize the importance of civil society's involvement in setting the disarmament agenda. It was heartening to hear many government representatives recognize the need for public awareness on the key issues of disarmament, arms control, human security and national security – and to recognize the role of NGOs in helping to build that public awareness.

At the same time, concrete reality reminds me that often the "partnership" between governments and civil society is fragile and tenuous and that even as some call for greater involvement of civil society, others are equally determined to turn back the clock not only on disarmament issues, but also on the involvement of civil society in shaping the terms of debate of disarmament – and many other issues, for that matter.

I am sure that many in this room will remember in July 2001, at negotiations at the UN to try to curb the proliferation of small arms and light weapons, the US delegation attacked the involvement of NGOs in disarmament issues, declaring that the US does "not support the promotion of international advocacy activity by international or nongovernmental organizations, particularly when those political or policy views advocated are not consistent with the views of all member states. What individual governments do in this regard is for them to decide, but we do not regard the international governmental support of particular political viewpoints to be consistent with democratic principles."

Certainly, the US is not alone in wishing to marginalize the voice of civil society, but to do so in the name of "democratic principles" confounds the mind – at least it confounds this mind.

Yet, this inverted notion of democracy is not inconsistent with the new lexicon of today's unipolar world. In this new lexicon, the definition of *multilateralism* reads "either you are with us or you are against us." In this new lexicon, *negotiation* means "you will accept my position, my framework, my world view, my bottom line – period." In this new lexicon, *peaceful exploration of space* includes military and intelligence-related activities. In this new lexicon, *realism* means "going along to get along" because if an individual or an entire country, for that matter, dares to question the underpinnings of this new reality, they are likely to be labeled "the enemy" at best and "evil" at worst.

A few days after the September 11 attack, I wrote a statement noting, "We do need to respond to the terrorist attack – but many worry what form the response might take. One US commentator reported that a former high government official speculated that perhaps even a nuclear response should be considered. Hopefully, more rational minds will prevail. How can anyone possibly think that nuclear weapons – the most indiscriminate, destructive weapon of all – are appropriate to consider as a response to a terrorist attack?"

When I mentioned my concern to a colleague, he was certain that either I had misheard the report or that it merely reflected the thinking of "a fringe element." Yet, with the new revelations of US nuclear policy – which have already been much discussed in this conference – how rapidly these words of concern have become clear reality in the post-September II world.

I share the feelings expressed by many over the past days. The challenges facing the global community are greater than at any moment since the end of the Cold War. Not only do we not want to lose the gains made in the 1990s, but we need to insure that there is a rational response to terrorism, firmly grounded in international law. The need to address terrorism is real. But how it is addressed is as critical to the outcome as is the need to deal with terrorism. We cannot allow the campaign against terrorism to mask the dramatically increasing militarism in the world that threatens to launch a new global arms race.

In order to deal with this impending crisis in disarmament, I believe that the role of civil society in framing and carrying out a disarmament agenda for the 21st century is more critical than ever. But I also share the view expressed by many here that governments too must take a more forceful role in shaping the world they want to see and not be held hostage to this moment in time.

Maj-Britt Theorin is one of the speakers who noted the dramatic changes that can be wrought by the involvement of civil society in disarmament issues, recalling the impact of mass European protest to nuclear weapons to be stationed there. Other speakers have noted that while the situation might look bleak at the moment, we all know that change is inevitable. That all might not be lost in what has been accomplished to date in arms control and disarmament.

In moving wildly, "schizophrenically," between despair about the present situation and "wild-eyed idealism" about the ability of the human spirit to rise above seeming impossible odds, I choose to focus on the inevitability of change and to be an active agent of the change I want to see. I choose to focus on what I believe to be the fundamental right of individuals and groups of individuals to determine their own future and not to have it determined for them in the name of "realism."

A few moments ago, I spoke about the movement to ban landmines, noting that it seemingly "came out of nowhere" to capture the public conscience and bring about dramatic change in a short period of time. But as anyone who knows the history of the movement to control certain conventional weapons can tell you, the "quick" success of the International Campaign to Ban Landmines and the resulting government-NGO partnership of the Ottawa Process was actually built upon much earlier work by the International Committee of the Red Cross, the UN and a few governments – such as Mexico and Sweden – that had begun calling for a ban of landmines – and other conventional weapons – in the 1970s. While that earlier arms-control effort had languished for over a decade when we launched the ICBL, it had served as a base from which we could launch our ban agenda.

Probably not surprisingly, I echo many of the points expressed at this conference by the Honorable Lloyd Axworthy – both in his various comments during the panels and in his discussion of the prevention of an arms race in outer space. Minister Axworthy notes Rebecca Johnson's call for a space focused "Ottawa Process" – a variation of the process that brought about the Mine Ban Treaty.

While it is widely held that civil society was the key to capturing the public conscience on the issue of landmines, the Canadian government was one of our key allies in forging the government–NGO partnership that made real change ultimately possible. And it was the bold personal leadership of Minister Axworthy in stepping outside the normal diplomatic mold to challenge the world to transform its mine ban rhetoric into treaty reality that gave our partnership a real framework in which to work.

We learned many things in the mine ban movement. And as the world now grapples with what a disarmament agenda for the 21st century might look like, it is incumbent upon those of us who believe that civil society has a meaningful role to play to work hard to not see the gains of the 1990s lost in the new "realism." Some of the relevant lessons[3] that we took from our work include:

3) These lessons and the elaboration that follows are developed from a chapter co-authored with Stephen Goose, for a book edited by Dr. Kenneth Rutherford of Southwest Missouri State University.

First, it is possible for NGOs to put an issue – even one with international security implications – on the international agenda, provoke urgent actions by governments and others, and serve as the ongoing driving force behind change. Civil society can indeed wield great power in the post-Cold War world.

Second, it is possible to achieve rapid success internationally through common and coordinated action by NGOs, like-minded governments, and other key actors such as UN agencies and the International Committee of the Red Cross. It is through concerted action that change is most likely to be effected.

Third, it is possible for small and medium size countries, acting in concert with civil society, to provide global leadership and achieve major diplomatic results, even in the face of opposition from bigger powers, and

Fourth, it is possible to work outside of traditional diplomatic forums, practices and methods and still achieve success multilaterally.

I would like to take a moment now to elaborate on some of these lessons, which might be applicable to critical issues facing us today:

- *Partnership Pays*

Perhaps the key factor in the success of the ban movement has been the close and effective cooperation between NGOs (primarily through the ICBL), governments, the ICRC, and UN agencies. This cooperation has been at the strategic and tactical level, and continues to this day.

- *Build a Core Group of Like-Minded Governments*

After being in an adversarial relationship with nearly every government from 1992–1995, an increasing number of governments began to endorse an immediate ban. The campaign called upon individual governments to come together in a self-identifying pro-ban bloc at the beginning of 1996, and they rapidly did so. Historically, NGOs and governments have too often seen each other as adversaries, not colleagues, and at first some in the NGO community worried that governments were going to "hijack" the process in order to undermine a ban. But a relationship of trust among a relatively small "core group" of governments (most notably Canada, Norway, Austria, and South Africa) and ICBL leadership quickly developed and has been maintained. This core group was geographically diverse, committed, and willing to provide leadership in the face of opposition from bigger states. The extraordinary dedication, energy, and tal-

ent of key governments and key individuals within those governments were absolutely vital in the success of the ban movement.

- *NGOs Need to Be Inside Too*

Obviously, a crucial role for NGOs is mobilizing public opinion.
But the mine ban movement has demonstrated that extensive involvement of NGOs in what are traditionally thought of as diplomatic activities is also crucial to insuring rapid and effective change. The
ICBL played a major role in the actual drafting of the ban treaty, from
its earliest stages. The ICBL was given a formal seat at the table in all
of the diplomatic meetings leading up to the negotiations, and then
during the negotiations as well. Not just plenary sessions, but all
working meetings were open to the ICBL. Many government representatives have since commented that the presence and input of the
ICBL and ICRC made a huge difference in the Oslo outcome. This
prominent and official role for the ICBL has continued during the
treaty intersessional work (with its Standing Committees of Experts
meetings) and during the annual meetings of States Parties of the
Mine Ban Treaty.

- *Non-traditional Diplomacy Can Work*

The Ottawa Process that led to the Mine Ban Treaty largely grew
out of the failure of negotiations in 1995–1996 on the Landmines
Protocol to the 1980 Convention on Conventional Weapons (CCW).
The pro-ban governments decided to pursue a "fast track" approach,
outside of traditional negotiating forums. The members of the Ottawa Process were self-identifying. Core group governments Austria,
Germany and Belgium hosted what were in essence preparatory
meetings, followed by negotiations hosted by Norway and a treaty-
signing conference hosted by Canada. Though done outside the formal UN structure, the Ottawa Process was strongly supported by various UN agencies, and especially by Secretary-General Kofi Annan.
The Secretary-General now serves as the depository of the convention.

- *Say No to Consensus*

From the start of the Ottawa Process, the core group governments
made clear that this was not to be a repeat of the Landmine Protocol
negotiations where any one country could thwart the will of others.
While always striving for consensus, the Ottawa Process stressed the
concept of like-minded, in essence saying, if you are not like-minded regarding a total ban, do not participate. During the Oslo negoti-

ations, the Rules of Procedure rejected the consensus approach, and while a vote never occurred, the fact that it took a two-thirds majority to make changes to the text contributed to the demise of several severely weakening amendments sought during the final negotiations of the Mine Ban Treaty in Oslo in 1997.

- *Promote Regional Diversity and Solidarity without Blocs*

Both the campaign and core group governments worked hard to ensure geographic diversity within the ban movement, and to promote a sense of ownership of the issue among regional organizations, especially the Organization of African Unity. This strategy paid off well. But parallel to that, it was important that the traditional diplomatic alignments never came into play; there were not "Western Group" or other regional meetings.

Certainly, not all of these lessons would be applicable in all cases, would be difficult to carry out in some cases, and could be counterproductive in others. Some have noted how the mine ban campaign had certain "advantages" – its focus on a single weapon, an easy to grasp message, its highly emotional content. Perhaps even more important, the weapon is obviously not vital militarily, nor important economically. But the difficulties encountered should also not be underestimated. There was virtually uniform opposition from governments at first, due to the widespread deployment of mines, considered by most as common and acceptable as bullets, an integral part of in place defenses and war plans, training, and doctrine. About 125 nations had stockpiles of antipersonnel mines; mines had been used in 88 countries. They were considered a cheap, low-tech, and reliable, substitute for manpower, but were also the focus for future research and development for richer nations.

These were long odds to overcome. But overcome them we did. We created the reality we wanted out of our vision of a world free of landmines. Those of us who believe that civil society has both the right and the responsibility to determine the 'reality' of our world must refocus our efforts and energies on expanding the dialog, interaction and partnership between NGOs, governments, international agencies and the United Nations – the body that provides a forum for all voices to be heard in developing a disarmament agenda for the 21st century. While sometimes the landmine ban movement worked outside traditional UN structures, it has also forged a strong and growing relationship with the United Nations. It is this UN, under the leadership of its Secretary General Mr. Kofi Annan that has worked diligently for broadening the involvement of civil society in its various deliberations. Witness, for example, this conference in Beijing. We can and must build upon these partnerships.

Perhaps these words sound like the mad rambling of an idealist. Of a person who does not have a firm and 'realistic' grasp on the state of the world today. Yet, I am not the only idealist in this room. Others at this conference – individuals from both the NGO world and government representatives – have confessed to being "idealists" if that means that we believe that we can create a world with a meaningful disarmament agenda that can encompass human security and national security.

Too often we "idealists" are told that the real world is a cold, hard, and unforgiving place and that to insure peace we must prepare for war. That is not a view that this idealist will accept. My view of realism is that you get what you prepare for. If we want to build a peaceful world, we must prepare for peace. If we want to live in a world with a meaningful agenda for disarmament in this century, civil society, like-minded governments, international agencies and the United Nations must forge a partnership to ensure that our "idealistic" vision becomes the new reality.

Jody Williams

Jody Williams, born in 1950, is a Political Scientist. She is the founder and coordinator of the International Campaign to Ban Landmines, an international Network of more than 1000 Non-Governmental Organisations. The Campaign and Jody Williams were awarded the 1997 Peace Nobel Prize for their engagement for the international convention to ban landmines.

Rapid Global Climate Change
– A Self-Made Imminent Threat for Entire Mankind

Hartmut Graßl

1. Climate as *the* Natural Resource

When we ask for the key climate parameters and the survival parameters for mankind we get nearly identical answers. For the first
these are: solar energy flux, precipitation, and surface properties especially of vegetation. For the second they are: again solar energy
flux, and water from the skies as well as photosynthesis of plants.
Rapid climate change must therefore impact deeply on all societies,
including the highly developed countries. The following example un-
derlines our dependence on climate: The major bread baskets of
mankind are humid to sub-humid temperate climates, in terms of
natural biomes called deciduous forests and steppe, where climate
and far less bedrock have led to deep soils with high carbon content
sustaining agriculture for many centuries, even at erosion rates
strongly exceeding soil formation rates.

2. Influencing Factors

Given the mean distance to the Sun and the size of the planet
Earth important climate factors are:

- spectral irradiance of the Sun
- quasi-periodic variations of the Earth's orbit around the Sun
- location of land masses
- internal interactions of climate system components
- volcanic eruptions
- impact of celestial bodies
- composition of the atmosphere
- human activities

Einstein – Peace Now! Reiner Braun and David Krieger (Eds.)
Copyright © 2005 WILEY-VCH Verlag GmbH & Co. KGaA, Weinheim
ISBN 3-527-40604-2

If we are only interested in the climate of the next few centuries we need not to consider the Earth orbital element variations as well as the changed location of continents, as these vary appreciably only on time-scales of several thousands to million years, respectively. However, we need estimates of the future luminosity of the Sun, a basic understanding of component interactions, especially between ocean and atmosphere, anticipated changes in atmospheric composition, mainly driven by industrialization, land-use changes through agriculture, settlements and industry. We cannot, at present, give estimates of future number, intensity and timing of volcanic eruptions as well as the impact of celestial bodies. While we know the global and continental-scale impacts of a given volcanic eruption, since we had two well-observed examples in 1982 (El Chichón, Mexico) and 1991 (Pinatubo, Philippines), there is not yet an observed impact of a celestial body on climate.

The major difficulty in assessing the importance of the different influencing factors lies in the shortness of observations for some of the factors. For example, the spectral irradiance of the Sun has only been measured since the late 1970s. During these two and a half 11-year solar activity cycles there was no trend observed for irradiance, but an oscillation of total irradiance with just below one per mille amplitude became obvious. Therefore, changes of the solar irradiance, often called solar constant, before 1978 have to be reconstructed from so-called proxy data, e.g. radiocarbon (^{14}C) content in tree rings depending on solar activity. The changes in atmospheric composition for the long-lived greenhouse gases carbon dioxide (CO_2), methane (CH_4), and nitrous oxide (N_2O) have to be derived from air bubbles in ice-cores recovered from Antarctica and Greenland, dating back now for four hundred thousand years. Paleo-climate data have thus become major tools when trying to project future climate change.

3. Delayed Reaction to Disturbances

Any system with interacting components, whose typical time-scales differ, must show a delayed response to any external disturbance, to which also our activities belong. The question that arises here is the delay of climate change at the surface caused by changed composition of the atmosphere. Since industrialization began in the late 18th century, human action has systematically increased the concentration of the three major long-lived greenhouse gases: CO_2 went up from about 280 parts per million by volume (ppmv), to now 375 ppmv with a five-year average growth rate of about 0.5 per cent per year at present; CH_4 has more than doubled from 0.7 to 1.75 ppmv, but shows a strongly slowed growth during the last decade; N_2O increased from 0.275 to 0.32 ppmv, at present increasing by 0.25 per cent per year. Many might ask why trace gases are so important? The answer is simple: major gases in the atmosphere like nitrogen (N_2) and oxygen (O_2) do not change radiation fluxes in the solar and terrestrial part as much as the trace gases mentioned. Including water vapor (H_2O) and ozone (O_3), the dominant and number 3 in the list of greenhouse gases, less than 3 per mille of the mass of the atmosphere determine to a large extent how solar radiation is absorbed and backscattered and how heat radiation is absorbed, scattered, and emitted in the atmosphere, including effects of liquid water and ice in clouds. As we can change only trace gas contents in the atmosphere appreciably, we nevertheless influence considerably the energy budget of the entire Earth. Taking only long-lived greenhouse gases this influence on the energy budget has surpassed one per cent of solar input, i.e. we have become a global climate factor.

How long is the delay between such a disturbance of the energy budget and the full reaction of the climate system? In other words: When will the warming caused by an enhanced greenhouse effect at a certain time have reached, say 80 per cent, under present continuing growth rates of the concentrations of greenhouse gases? Three to four decades later. Hence, the observations of today cannot possibly show the full reaction. Would we be able to stabilize concentrations of greenhouse gases, the goal written into the United Nations Framework Convention on Climate Change (UNFCCC), the warming would continue at a slower pace for decades and sea level rise would not stop for centuries or even millennia, as very slowly reacting system components like the deep ocean and ice sheets would still be in the adaptation process.

Mankind has started a global experiment whose outcome at the regional scale is highly uncertain and that will take at least centuries before natural, less rapid climate change will be dominant again.

4. Has Climate Change already been Observed?

Climate can never be stable, because system components with strongly differing time-scales interact; creating what is called natural variability on time-scales smaller than the reaction time of the slow component. In addition, external factors like solar irradiance vary as well on different time-scales. Hence the difficulty to distinguish an observed trend of the last hundred years from a longer-term internal oscillation, which is part of natural variability. Before I try to attribute observed changes to causes, the mere facts of observed climate change during the instrumental period with direct measurements of climate parameters are given in a condensed manner [1]:

a) Mean near surface air temperature has risen in the 20th century by 0.6°C, and the warming is above average in inner continental areas;

b) Precipitation has increased in high northern latitudes throughout the year, in mid-latitudes winter precipitation increased in most areas;

c) Mountain glacier retreat is nearly global with a recent acceleration;

d) Mean sea level has risen by about 15 cm in the 20th century, accelerated to 3 mm per year since 1992, when consistent satellite observations became available on a global scale;

e) Sea ice extent in the Arctic shrank by about 3 per cent per decade since 1972 in late winter and by about 6 per cent per decade in late summer. No significant sea ice trend is observed in the ocean around the Antarctic;

f) Air pollution has a strong impact on reflectivity of clouds as derived in areas with major changes in air pollution since 1981 from satellite data, e.g. Central Europe and China;

g) Precipitation amount per event has increased, in most areas with slightly less, equal or increasing total precipitation.

Are all these changes anthropogenic? Yes, to a large extent, as already expressed in the following sentence: *The balance of evidence suggests a discernible human influence on global climate* [2].

To reach this conclusion the observations reported above were compared with the history of the climate forcing (greenhouse gases, solar irradiance, and air pollution) and model calculations applying the forcing and reproducing the observed 20th-century variability and changes. In addition, statistical techniques to find emerging anthropogenic climate change patterns in a "noisy" record were employed. Despite this attribution of climate change to several anthropogenic causes by several groups and the assessment of their publications by the Intergovernmental Panel on Climate Change (IPCC) many media will still attach large question marks to the above statement, as we embark into the first climate protection measures that would see losers and winners in industry and commerce. The losers will try to blow up uncertainties to get a further few years delay before measures enforced by international law are introduced.

5. What Scenarios of Future Climate Might be Probable?

The large range of global mean warming (1.4 to 5.8°C) given in 2001 by the IPCC Working Group I (The Science of Climate Change [3]) when climate modellers had used all IPCC scenarios was very often interpreted as the uncertainty in climate modelling. However, it was to a large extent due to the spread of the scenarios with respect to the degree of multilateralism, economic growth, technological innovation and environmental concern (not climate protection). Hence, the large spread was caused by our inability to select a probable scenario of the development of mankind in the 21st century. Would mankind select a joint goal and try to reach it by international law within the United Nations Framework Convention on Climate Change (UNFCCC) the situation would change drastically, as a probable scenario would have been defined, given compliance with international law. The Global Change Advisory Council of the German Federal Government (WBGU [4]) has developed and proposed such a scenario built on the following basic assumptions: Maximum +2°C mean global warming in the 21st century, multilateralism wins (UNFCCC will be implemented), mankind fosters innovative technologies, successful development of developing countries (high economic growth scenarios). The main pillars of such a development are: renewable energy sources become dominating, gross national product reduction for the restructuring of the energy systems remains well below 2% in all regions (i.e. much smaller than the estimated damage and adaptation costs without climate protection), acceptance of the equal right to emit for any person (adoption of contraction and convergence). The key political hurdle in this context is the rapid integration of emerging countries, like China and India, into the countries with commitments for a less-polluting development path. This full integration has to start for the commitments after the Kyoto Protocol (beginning in 2013). Because many countries set goals in this direction at the "Renewables 2004" in Bonn, Germany, the probability that a path proposed by WBGU will be followed has grown.

6. Which Climate Change Impacts will be Highly Probable

There are several obvious impacts of climate change caused by an enhanced greenhouse effect: Accelerated mean sea level rise, redistribution of precipitation, changed frequency distributions of all climate parameters at any location. The latter implies, as the tails of frequency distributions contain the extremes, more frequent weather known extremes *and* new weather related extremes. We have many indications for broadened but not for narrowed frequency distributions, except for the daily temperature amplitude over many continental areas. Therefore, all our security infrastructure, built according to hitherto observed climate variability, is no longer fully adapted and has to be changed, if the same security level has to be reached again.

Can we already point to such climate change impacts? As already indicated in section 4 the amount of precipitation per event has increased in areas with no increase or even a light decrease of total precipitation, i.e. flash flood events became more frequent but also dry spells often longer. This physically plausible result follows from the strong increase of water vapor saturation concentration with temperature (~8 per cent increase per 1°C warming).

Another observation is coastal erosion at most coastlines, pointing to a sea-level rise rate surmounting sedimentation rate. Many more abnormal changes, like invasion of new pests or coral bleaching have been observed but due to a mix of global and local change the causes cannot be clearly separated.

Because subsistence farming communities are especially vulnerable to climate change, global anthropogenic climate change causes a major ethical problem: the real polluters in the industrialized or industrializing world will suffer much less than those not causing the changed atmospheric composition. Therefore, any climate-protection policy has to include climate change adaptation costs for the least developed countries, as already foreseen in the Kyoto Protocol's Marrakech accords.

7. Energy Systems as the Decisive Factor

The two largest challenges of the world community are rapid development of developing countries and the rapid approach to sustainability for developed and developing countries, i.e. we need a higher energy throughout at climate protection. In other words: Tripling primary energy input at much less emissions until about 2050, even when assuming a major energy efficiency increase. Referring to Section 5 this is only feasible if multilateralism prevails, if external costs are internalized, i.e. subsidies for fossil fuels of about 300 billion € per year are reduced rapidly and agricultural protection measures (again close to 300 billion € per year) are relaxed.

In order to keep the environmental load small when multiplying the renewable energy share needed for the transition towards a sustainable energy system, the direct solar energy must get the lion's share in the long run. A small comparison of the energy flux densities will make this clear: The Sun delivers about 170 watts per square meter ($W m^{-2}$) as global mean downward irradiance of which the biosphere transforms – on global average – only 0.1 $W m^{-2}$ into biomass. Thus a major research push for the more efficient and economic solar energy use in the coming decades is more important than development of so-called energy forest for the energy input to all our activities, which have nearly reached 0.03 $W m^{-2}$ on global average.

Our energy system would no longer be the key environmental disturbance, if less than 0.1 $W m^{-2}$ from 170 $W m^{-2}$ are used. However, if we wait with the transition certainly adaptation but also mitigation costs would skyrocket. Energy research must regain its importance, whereby technological progress is just one of several components. Also drastic increases of energy productivity through reduced material fluxes and demand-side management are of similar relevance.

8. A Strategy for Sustainable Energy Systems

Firstly, the strong, industrialized countries need to make a sustainable global energy system a top agenda item. In this context I welcome that partial debt release for developing countries not just the least developed countries and climate protection are for the first time high up on the agenda of the forthcoming G8 summit in Great Britain in 2005. Hopefully, the connection of both topics to sustainable energy systems for all countries will become obvious during the summit.

Secondly, we must establish a far-reaching goal to initiate transition into sustainable energy systems. As development of poor and therefore vulnerable countries is also an ethical issue such a goal must have as its base the equal right of each individual to emit; in other words: the polluter pays principle is an international and not only a national one. Therefore, contraction and convergence of emissions must lead in a certain time frame to similar globally reduced emissions per capita. The Global Change Council of the German Federal Government has proposed [5] to reach this goal in about half a century. It would also mean a strong incentive to reduce fossil energy subsidies in all countries.

Thirdly, the negotiations to go beyond the Kyoto Protocol, which have to start officially at the 11th Conference of the Parties to UNFCCC in 2005, if unable to agree on the above strategic goal, need – as a minimum – to enlarge the list of countries with commitments, whereby the group of emerging industrializing countries must get offers from the OECD countries to start sustainable energy systems, for example by energy partnerships.

An energy system based to a large extent on renewable energy would give each country much less dependence from concentrated energy resources like oil and thus would offer a major peace dividend. It would also dampen the anthropogenic climate change rate within several decades and allow to reach the UNFCCC goal of stabilized greenhouse gas concentrations.

References

[1] For more details see: *Science of Climate Change*, Report of Working Group I of IPCC, Cambridge University Press, Cambridge, UK, 2001.

[2] See IPCC 1996.

[3] See IPCC 2001.

[4] Download of all reports from www.wbgu.de.

[5] See: www.wbgu.de, *Special Report Beyond Kyoto: Sustainable Climate Protection Strategies for the 21st Century.*

Hartmut Graßl

Hartmut Grassl was born in 1940. Since 1988 this Physicist has been the Director of the Max-Planck-Institute for Meteorology in Hamburg, Germany. From 1994 until 1999 he was Director of the UN Project "World Climate Research Programme".

Part 4

Finding Peace in the Middle East

"One of these ideals is peace, based on understanding and self-restraint, and not on violence. ... our relations with the Arabs are far from this ideal at the present time".

Albert Einstein

Einstein – Peace Now! Reiner Braun and David Krieger (Eds.)
Copyright © 2005 WILEY-VCH Verlag GmbH & Co. KGaA, Weinheim
ISBN 3-527-40604-2

Peace in the Middle East: A Global Challenge and a Human Imperative

Hanan Ashrawi

It is precisely during such times of adversity and pain, of violence and victimization, of unilateralism and militarism, of ideological fundamentalism and absolutist exclusivity, that the world is most in need of voices and forces of sanity, reason and moral responsibility – the genuine building blocks of peace. As we witness attempts at imposing a simplistic view of a Manichean universe, of polarization and reductive stereotypes of good and evil, we are most in need of those who will engage in a redemptive validation of pluralism, tolerance, diversity, authenticity of identity, and the comprehensive engagement in collective responsibility. As such, it is up to us jointly to give both a voice and an audience to the silenced, and to grant space and time to the excluded and denied.

Such is the nature of intervention that the world requires, not only to resolve conflicts but also to prevent them from erupting or generating their own destructive forces that could spiral out of control. No conflict should take us by surprise, for all the symptoms are recognizable and the components definable. Long-standing grievances and inequities have become all too familiar and have been left to fester on their own or to be manipulated by the strong as a means of victimizing the weak. The nature of pre-emptive action must be, by necessity and choice, constructive, peaceful, and therapeutic.

Since an aspect of globalization is the redefinition of enemies and allies, friends and foes, crossing national, territorial, and cultural boundaries, the process of rectification must also utilize the means made available by the knowledge and IT revolution as tools of contemporary global realities. Thus hunger, poverty, illiteracy, the spread of disease, the degradation of the environment, the disenfranchisement of the weak, the suspension of human rights, among others, are all universal enemies that require the collective effort of universal allies. Human-based development programs and inclusive sys-

Einstein – Peace Now! Reiner Braun and David Krieger (Eds.)
Copyright © 2005 WILEY-VCH Verlag GmbH & Co. KGaA, Weinheim
ISBN 3-527-40604-2

tems of governance remain the most appropriate means of empowerment.

Most significantly, the indispensable universal instruments remain those that ascertain a global rule of law, encompassing both state and non-state actors, capable of assessing culpability, providing accountability, and ensuring redress with justice. Along with their multilateral institutions, they remain safeguards against unilateral power on the rampage or destructive military pre-emption on the basis of subjective criteria.

With that in mind, peace in the Middle East, or the just solution of the Palestinian–Israeli conflict, can be addressed in its proper context as the longest standing case of military occupation and as the most persistent unresolved case of denial, dispossession and exile in contemporary history. As such, it is also an anachronism in that it has all the components of a colonial condition in a post-neocolonial world, plus the requirements of national self-determination as a basis of nascent statehood in a world moving towards regional and global redefinitions.

Regionally, the conflict has provided a convenient excuse for the suspension of human rights, the evasion of democratic systems of governance, the waste of natural and human resources, and the perpetuation of centralized regimes that held back the challenges of development – all under the guise of 'national security' and the external military threat. For decades, war, or the threat of military hostilities, has served to maintain the status quo and has framed the region within misplaced notions of self-defense that contributed to the rising power of extremism and fundamentalism rather than human empowerment and global engagement.

Peace, or the prospect thereof, is possibly the most effective force for dislodging such notions and trends, becoming, de facto, the most destabilizing factor in a region suffering from an imposed state of suspended animation. The legacy of colonialism clearly has served the interests of those in power, predominantly client regimes, who sought to maintain control, thereby leading to the collusion of internal and external forces in the exclusion of the people as a whole. A just and comprehensive solution to the Palestinian-Israeli (and hence Arab-Israeli) conflict would unleash all those forces so far held in abeyance, but forming the indispensable energy for sustainable progress, development, democratization, and regional integration.

While threatening short-term stability based on restrictive and constrictive norms and patterns, it constitutes the sole mechanism for any stability that can lay claim to permanence on the basis of contemporary and future-oriented political, social, cultural, and economic systems of cooperation and interdependence.

Globally, the Palestinian question remains central to any human vision of globalization as a test of the collective will to intervene and to maintain a global rule of law based on operative principles of justice and historical redemption. Granted, the current dynamic is antithetical to the aspirations of peacemakers who had based their endeavors on the universality of human rights, parity before the law, positive intervention, and the non-violent resolution of conflicts through redress and the elimination of grievances. A serious paradigm shift is necessary for the restoration of these human values that have been subverted in the aftermath of September 11 and the triumph of the neoconservatives and fundamentalist ideologues in key power centers.

The logic of peace that had been formulated painstakingly (and painfully) as the substance of Palestinian–Israeli encounters and dialogs, even long before negotiations, is currently being drowned by the din of war drums and the frenzied mutual infliction of pain over the last three years. Such tragic and unprecedented pervasive violence is not only eradicating previous achievements and agreements, but is also destroying the prospects of any future reconciliation. Its most alarming impact is on the perceptions and attitudes of both peoples, particularly in the regression towards the fallacies of the past and the stance of mutual negation emanating from the revival of deep-seated existential fears of survival.

Such fallacies and false assumptions must be boldly confronted and systematically deconstructed if there is any hope of extricating both sides from this lethal and self-perpetuating trap of mutual destruction.

The notion that a whole nation can be brought to its knees by the use of unbridled violence, or that the will of a people can be defeated by military means must be discarded once and for all. Armies may be able to defeat other armies, but the limits of power are most apparent when used against civilians and non-combatants. Along with that, the fallacy that there is or can be a military solution to the conflict must be completely and irrevocably discarded.

Conversely, the emergence of the bizarre concept of a 'balance of terror' has reinforced the irrational and immoral killing of civilians and the victimization of the innocent. The drive for revenge, like the escalation of military brutality, has generated the most tragic and futile momentum for escalation and self-destruction. On both sides, the 'no holds barred' mindset has taken over as a mindless, visceral, repetitive response with horrific ramifications. The erroneous assumption that greater pain and punishment, or the escalation of failed measures, would somehow lead to 'success' or the surrender of one side to the other is at the heart of the prevailing dynamic of death and devastation.

Related to that is the notion that a people under occupation will eventually come to be reconciled to the fact of their captivity and to accept their fate without struggling for freedom and dignity. Self-determination to the Palestinian people is not an abstraction, but the actual realization and enactment of their identity on their own land, and a motivating force for independence and statehood. It is the final negation of the myth of a 'land without a people for a people without a land' that has long framed the rationalization for the most extreme forms of Zionism that sought to deny the very existence and humanity of the Palestinians.

For the conflict to be resolved, its causes must be identified and solved, while grievances and fears on both sides must be addressed and laid to rest. Neither side can lay claim to a monopoly of pain and suffering, in the same way as it cannot claim exclusivity of narrative and legitimacy. Clearly, peace cannot be made incumbent upon converting all Palestinians to Zionism or transforming all Israelis to espouse Palestinian nationalism.

The denial or distortion of the narrative of the other has served as a convenient vehicle for the dehumanization of the adversary and hence as a justification for all forms of violations and atrocities while evading accountability. Historical records must be reconciled, whether in the recognition of the horror of the holocaust and all its horrendous implications, or in the historical victimization of the Palestinian people and their dual tragedy of dispossession and exile, on the one hand, and oppression and occupation on the other.

It should also become apparent that, ironically, in this context the Palestinians and Israelis have reached the stage of dependent legitimacies rather than a competition over a singular and mutually exclusive legitimacy. Since the essential requirement for peace lies in sharing the land of historical Palestine, it follows that there has to be a shared legitimacy based on parity and mutuality. Neither side can (or should be allowed to) destroy the other physically, morally, or legally. A full admission of equal value to human lives and rights must be internalized, with no claims to superiority on those most essential human values and attributes.

In the same way, there can be no exclusivity of claims – whether to the land or to security or to the discourse and public presentation of the issues. Shared boundaries exist both as territorial and as moral/human concepts of proximity and interaction. Security, therefore, is a factor of mutuality and interdependence, emanating from the core considerations of the totality of human imperatives. Historical, territorial, cultural, economic, social, personal, existential, legal, and political dimensions of security must shape the issues and drive the process beyond the narrow confines of military security. A human and humanistic strategic approach to peace is by definition one of integrated empowerment rather than the stratagems of power politics or coercion or military control.

At the opposite pole, the fallacy of fundamentalism, or even divine intervention and dispensation, has been exploited to justify absolutism and exclusivity, thereby ending all hope of a solution based on accommodation, while claiming unrestricted license to kill and destroy. Extremist ideologies tend to thrive in times of despair and insecurity, and like the recourse to violence and militarism, they signal an absence of effective workable solutions and handles on reality.

Radicalization is also a factor of distortion in the sweeping ideologies and simplistic generalizations of theories such as the 'clash of civilizations' or 'war among religions' or the imposition of democracy by force of arms. Increasing polarization widens the gap and warps any vision of reconciliation not only by depicting the conflict as part of a grand sweep of teleological proportions, but primarily by rendering it impossible to resolve through available peaceful means of practical and legal disentanglement. Inevitability of conflict as defined by an abstract universal design is directly antithetical to responsibility and intervention.

By now it has become apparent that the assumption that the Palestinian-Israeli conflict is a purely bilateral issue and can be resolved by the two sides without third-party intervention is entirely false. It has been variously used to maintain the asymmetry of power, to justify the lack of political will or the impotence of external actors, and to sustain other false assumptions such as the 'peace through exhaustion' fallacy or 'intervention following sufficient bloodshed'.

The need for third-party intervention is not only a factor of balance, but an indispensable force for breaking the lethal cycle of violence and revenge, while providing a context for legality, arbitration, and guarantees. A genuine form of multilateralism and collective responsibility is the sine qua non of the resolution of this conflict. Artificial, unilateral, and power separation such as that represented by the expansionist apartheid wall is a recipe for further conflict and greater violence – not least for encapsulating many forms of coercive injustice including land and water theft, fragmentation of Palestinian reality and the creation of isolated ghettoes, and imposing political boundaries that destroy the chances of a viable Palestinian state, hence of a just peace.

Palestinian nation-building and statehood are imperative for peace and stability throughout the region. Democracy and separation of powers, the rule of law and respect for human rights, institution-building and good governance, transparent accountability and reform – all are the ingredients of viable Palestinian statehood. The occupation, however intrusive, must not be used as an excuse to avoid responsibility. Similarly, negotiations and compliance with agreements must not be suspended pending the establishment of a Palestinian Utopia. Devolution of occupation and evolution of statehood must proceed simultaneously with urgency and commitment as interdependent processes.

An instrument like the Road Map of the erstwhile Quartet could have served as a lifeline for peace had it been implemented with speed and integrity, with clear timelines, monitoring and verification mechanisms, and the courage to exercise impartial accountability. The incorporation of the Israeli amendments in the implementation has tarnished the integrity of the text and of the external actors as well. Frontloading the process with Palestinian obligations, adopting the sequential and conditional approach, and creating further interim phases without guarantees on the ground have rendered the Road Map inoperative and subject to extremists on both sides. Absent political will, even-handedness, and seriousness of intent, third-party intervention could backfire and aggravate the conflict further through dashed hopes and let-downs.

However, third-party interventions can also be destructive if motivated by special agendas, if they exercise bias, and if they are incapable of effecting reality on the ground. Without substance, legitimacy, and applicability such interventions create a semblance of engagement without coming to grips with the reality of the conflict itself. When the issue is relocated domestically to become part of internal political realities, particularly in election votes and funds or the influence of special interest groups, then the question becomes one of exploitation and self-interest rather than serving the cause of peace.

The most detrimental external interference is that of the zealots and enthusiasts who embrace the most extreme long-distance stances with the 'passionate intensity' of the 'worst.' Blind loyalty for, and identification with, one side lead to the adoption of the most strident belligerency towards the other, hence intensifying the conflict and subverting dialog and rational communication. Islamic fundamentalists and regressive brands of Arab nationalists have ironically joined forces with Christian evangelicals, Jewish fundamentalists, and ideological neoconservatives to fight their own proxy wars at the expense of moderate Palestinians and Israelis alike. Such radical apologists have inflicted serious damage and pain from their safe distance in Riyadh, Damascus, Washington, Knoxville, or Sydney demonstrating the type of intervention that no peace can survive.

The superimposition of blind loyalty or guilt has revived the worst of racist labeling and dehumanization with the additional superimposition of false analogies. It may be convenient to label all Palestinians as 'terrorists' and dismiss them from the conscience of the world in the context of the 'war on terrorism.' It may be equally convenient to describe the Israeli occupation's measures of aerial bombardment and shelling of Palestinian civilian areas, of assassinations and abduction, of home demolition and destruction of crops, of siege and fragmentation, of checkpoints and humiliation, of illegal settlements and apartheid walls and annexation fences as legitimate forms of 'self-defense.' It may be comfortable to dismiss decades of military occupation and dispossession as figments of the victim's imagination, hence irrelevant to the current conflict. However, such scoring of points only makes the solution all that more distant.

So far, the solution remains simple and attainable, having been repeatedly defined and having become part of a global consensus. The two-state solution is still possible, though becoming increasingly more difficult with the expansion of settlements, by-pass roads, and the apartheid wall throughout Palestinian territory. The bi-national state as a de facto solution will become the only option should Israel continue its expansion and its refusal to withdraw to the June 4, 1967 lines and remove the settlements of the West Bank and the Gaza Strip. Territoriality will give way to demography, and the issue then will become one of democracy, with Zionism forced to re-examine its most basic premises.

Jerusalem, both East and West, can become an open city and the shared capital of two states, thus encapsulating the essence of peace and regaining its stature as a city much greater than itself and not subject to exclusive possession or greed of acquisition. The Palestinian refugees must be granted historical, legal, moral, and human recognition and redress in accordance with international law and the requirements of justice. There is no need to reinvent the wheel, but there is a need for the will and courage to act against all adverse forces.

Dear friends, sisters and brothers – as we hurtle towards the abyss, as we daily lose unique, irreplaceable lives, and as attitudes and hearts are hardening, may I take a moment to recognize this luminous instant in history that you are affording us. You have chosen to intervene on the side of those who have decided to take risks for peace rather than those who thrive on hate and conflict. It certainly takes a unique form of courage, tenacity, and distinctive human priorities to challenge prevailing fallacies and injustices. On behalf of the Palestinian people as a whole, and on behalf of all Palestinians and Israelis who have maintained their partnership for peace, and on behalf of all those who are in solidarity with our joint effort, I thank you. You have taken up a global challenge, and you certainly embody its human dimension. We are indeed heartened and empowered.

Hanan Ashrawi

born in 1946 in Ramallah, West Bank, studied at the American University of Beirut (masters degree) and at the Virginia State University, Charlottesville (Promotion to Dr. Phil.). She acted as official spokesman for the Palestinian delegation of the Middle East Peace Conference in 1991 in Madrid. In 2003 she was awarded the Sydney Peace Prize.

Einstein's Legacy
Felicia Langer

Albert Einstein died on April 18, 1955, a week after signing Bertrand Russell's peace manifesto. One of the most significant points of the manifesto stated:

> In view of the fact that in any future world war nuclear weapons will certainly be employed, and that such weapons threaten the continued existence of mankind, we urge the Governments of the world to realize, and to acknowledge publicly, that their purpose cannot be furthered by a world war, and we urge them, consequently, to find peaceful means for the settlement of all matters of dispute between them. [1]

The signing of the manifesto was Einstein's final political act – and my first in Israel during the early 1950s. With others, I collected signatures to petition against nuclear war. So far away from Einstein, and yet so close. He inspired millions through his rigorous fight for peace, until his final breath.

In 1954, Einstein wrote about Israel, a country for which he held deep affection:

> The Jewish people who themselves have suffered so severely from prejudice and oppression must have complete understanding for the necessity to grant the Arab minority in Israel freedom, democracy and full equality. [2]

Einstein was not fully aware of the injustice that Israel had inflicted upon the Palestinian people. In 1950 I saw with my own eyes the demolished Arab villages and heard the stories of the expulsion and massacre of 1948. I experienced the expropriation of the Palestinian land and the discrimination of Palestinians as second-class citizens who did not enjoy the same freedom and democracy, not to mention equality.

Einstein – Peace Now! Reiner Braun and David Krieger (Eds.)
Copyright © 2005 WILEY-VCH Verlag GmbH & Co. KGaA, Weinheim
ISBN 3-527-40604-2

"... our conduct towards our Arab minority is the true test of our moral standards," wrote Einstein further on this subject. Had he come to know the condition of the Arabs in Israel back then, the mark he awarded for our moral standard would have been very poor, and even worse for the current situation.

The Six-Day War was followed by years of illegal occupation of the Palestinian territories that has now lasted more than 37 years. I have experienced the years of occupation as a contemporary and eye-witness and seen the whole extent of oppression: the expulsion of the Palestinians, the dispossession of land, and the illegal settlements as well as the massive destruction of houses. The worst part was that I have seen the wounds of tortured Palestinians. Yet, some I was unable to see because they had been tortured to death.

The Israeli peace movement can attest to how the Israeli government rejected with condescension and contempt the peace initiatives of Nahum Goldmann, Mendes France, Bruno Kreisky and others for making peace with the Palestinians and the Arab world. During the years from 1991 to 2000, we held peace negotiations with the Palestinians; the OSLO Peace Accord (between the PLO and Israel) was signed during the years from 1993 and 1995. It was an arduous peace process, but without peace. All Israeli governments (also that of Rabin–Peres) have further settled the occupied territories and created a fait accompli. By the year 2000, the number of settlers had doubled (from 92,000 to 200,000 in 2000, and an additional 180,000 in East Jerusalem). The Prime Minister at the time, Barak, was called the greatest settlement builder, and he allegedly wanted to make "generous peace" with Palestinians...

How could that happen? Shimon Peres in 1996 gave a revealing response: Because the Palestinians were so weak, they became a suitable partner for us. And in 1993, 12 days before the signing of the OSLO Accord on 13 September 1993, Peres determined Israel did not negotiate with the PLO, but with a shadow of itself...

The consequence was that the Palestinians' most important characteristic making them suitable as negotiating partners for us was their weakness... As has been true for centuries, the maxim still holds "Vae Victis – Woe to the conquered!" – This concept of peace has led to the fact that there is no peace, that the occupied territories have become a scene of Apartheid where the settlers enjoy a "master race democracy" and the Palestinians live under oppressive occupation. This "virtual peace" exacerbated the Palestinian's situation and led to the start of the second intifada. The cruel suppression of the intifada, ab initio, when it was still unarmed, when according to sources from the Israeli military 1 million shots were fired at Palestinians (a bullet for every child, one cynical slogan went), led to an escalation in the spiral of violence. These and other measures that I will mention have so stirred up and frustrated Palestinians that seeds were sown in Israel for the murderous suicide bombers to which hundreds of innocent Israelis have fallen victim.

Avraham Burg, a leading member of the Labour Party and former head of the Knesset, wrote:

> If Israel has grown indifferent to the children of the Palestinians, it should not be surprised when full of hate they blow those places sky high where the Israelis entertain themselves so as not to have to look reality in the eye. They dare to go to the places where we spend our leisure time, Allah an, because their life is a torture. They spill their blood in our restaurants to ruin our appetites because at home they have children and parents who endure hunger and are humiliated. [3]

We have developed an Orwellian euphemistic language: Extra-judiciary executions of Palestinian suspects and of those in their vicinity who happen to be present are called "targeted killings". The military invasion of the West Bank whose destructive force was primarily directed at the civilian population and public buildings was called "Operation Protective Wall". The term "Dividing structure for security purposes" was given to a wall that for stretches is being built deep into the Palestinian territory resulting in a massive expropriation of land and the uprooting of fruit and olive trees (an estimated 100,000 trees) and has cut off innumerable Palestinians from their land, their water sources, schools, workplaces, and hospitals.

This would all not be possible if it weren't for the massive, unconditional support on the part of the USA. The ingenious weapons that are implemented against the Palestinian civilian population, the bulldozers that destroy their houses are either "American made" or "American paid". Not to mention the nearly blind political support that America exerts by making use of its right to veto in the UN Security Council in favor of Israel.

What would Einstein have said to this? In 1952, he said, although back then the circumstances were quite different: "Our good America is striving successfully to be a caricature of itself."

"No individual has the right to call himself a Christian or Jew if he is prepared to murder according to plan by order of an authority." Einstein wrote in 1928.

In Israeli Palestine, the situation is thus: Approximately 2 years ago, the Israeli airforce dropped a 1-ton bomb on a house in Gaza as part of a so-called "targeted killing" in order to kill a suspected Hamas activist. This bomb killed 15 innocent bystanders, including women and children. Such "collateral damage" (the term was voted nonsensical word of the year) was accepted and has occurred frequently. The commander of the airforce at the time and recently named commander-in-chief of the army Dan Chalutz commented that he could sleep well with the results of the bombing... but not some of the Israeli pilots who have to carry out these kinds of criminal orders. They demonstratively declared that they cannot sleep well with it and that they refuse to participate in the bombing of civilians. This was the first time in the history of Israel that pilots, the elite of the Israeli army and society became "refuseniks". ...There are many young Israelis now who conscientiously object to military service in the occupied territories because they do not want to be involved in the war crime activity perpetrated there by the army.

In 1953 Einstein wrote on this topic, "There is still one human right – the right or the duty to exclude oneself from participating in activities if one perceives them to be unjust or corrupt."

Our peace movement, human rights organizations, "Women in black", our best daughters and sons are our hope. They are the other face of Israel. They build the bridges of peace to the oppressed Palestinians. Both need international solidarity.

The Palestinians in their overwhelming majority have made their willingness for peace very clear, and indeed they have done so for years. Not even Arafat was an impediment to peace as the infamous Israeli-American legend claimed. Arafat was not the problem, the occupation that continues today is the central problem.

The elections on January 9, 2005 confirmed this (completely democratic elections, by the way, even though they were carried out under oppressive occupation). The Palestinians are for a two-state solution, for a sovereign, viable Palestinian state at the side of Israel in the territories occupied since 1967, and for a just resolution of the refugee question in accordance with international law. It is a clear preparedness for compromise on the part of the Palestinians who are satisfied with only 22% of historic Palestine.

We would have to clear out of the occupied territories, including the settlements, in accordance with international law. We would have to follow the old Resolution 242 of the International Security Council from 1967 that is gathering dust in the archives. This resolution clearly states that the acquisition of land by means of war is prohibited. Until today, each Israeli government, including the government of Sharon (with and without Peres) has rejected this resolution. The clearing of all occupied territories, not just Gaza and not just partial clearing, is the conditio sine qua non for peace with justice that will also guarantee Israel's security. Sharon's phrasing on peace is not convincing since it is not followed by deeds.

He speaks of peace and at the same time Palestinian land is expropriated to build the wall that the highest court of the UN has declared to be illegal in its current and planned course, in a vote of 14 to 1. Likewise the General Assembly of the UN declared it illegal with a vote of 150 to 6. Unfortunately the International Security Council is paralysed by the veto of the USA, nonetheless the international community has condemned the Israeli measures as expressed in Den Haag and in the General Assembly.

To achieve peace with justice, the catastrophic Israeli politics based on the arrogance of power would have to change. To achieve this, international pressure would have to be exerted on Israel and the long-term violations of human rights denounced. This pressure in the name of peace will be a blessing for the Israeli people for the politics of the previous governments are not just a tragedy for the Palestinians, they are a tragedy for the Israelis as well.

It would be of the same mind as Einstein, in the sense of his willingness to stand up for what he believes and his sincerity. In 1931 he wrote in a letter to a friend in England: "Whatever I do or say cannot change the structure of the universe in the least. But perhaps my voice can help promote the greatest of all ideals: good will among men and peace on earth."

To raise one's voice for peace with justice is imperative for individuals with a conscience everywhere.

That too is Einstein's legacy.

References

[1] Albert Einstein, *Über den Frieden*,
Abraham Melzer Verlag, Neu-Isen-
burg, 2004.

[2] Ibid.

[3] TACHELES, 11 Sept 2003.

Felicia Langer

born 1930 in Tarnow, Poland, as Child of jewish parents, escaped the Sowjet Union's Nazis in 1939 and returned to Poland after the war. In 1950 she emigrated to Israel, studied jurisprudence and worked deeply committed from 1965 as human rights lawyer for Palestinians. Rewards: Alternative Nobel Prize (1990), Bruno Kreisky Prize for Human Rights (1991), Erich-Mühsam-Prize (2005). Felicia Langer is the author of numerous books about the Israel-Arabian conflict.

Part 5
The Responsibility of Scientists

"In our times scientists and engineers carry particular moral responsibility, because the development of military means of mass destruction is within their sphere of activity."

Albert Einstein

Einstein – Peace Now! Reiner Braun and David Krieger (Eds.)
Copyright © 2005 WILEY-VCH Verlag GmbH & Co. KGaA, Weinheim
ISBN 3-527-40604-2

Remembering Einstein : Science, Ethics and Peace

Ronald McCoy[1]

Introduction

Albert Einstein, the renowned physicist, Nobel Laureate for science, and peace activist, was born in Germany on 14th March 1879. He is known for the general theory of relativity that altered the world's conception of space, time and energy. In his efforts to understand the universe, Einstein laid the theoretical foundations for nuclear energy, which eventually led to the development of the atomic bomb that has given a radically new dimension to war and security.

Science can be traced to its early beginnings in Islamic civilization between the ninth and thirteenth centuries, a time when Christian popes were burning 'witches' in the gloom of the Dark Ages. After the thirteenth century, when Islamic civilization suffered a severe regression, it was Western civilization that gained ascendancy. About five hundred years ago, learning in Christian Europe was deeply rooted in the pre-scientific philosophy of Aristotle and strongly influenced by the divine authority of the Church. Then came the cultural upheavals of the Renaissance and the Enlightenment, which swept away the medieval world order, shattered the temporal authority of the Church, rejected feudalism, sowed the seeds of humanism and capitalism, and gave birth to modern science in Europe.

At the heart of the Scientific Revolution of the seventeenth century was the glorification of humanity and human reason. It built upon the earlier scientific advances of the printing press, the telescope, the microscope and the vacuum pump, and transformed the intellectual and scientific landscape. The Scientific Revolution also received a boost from Copernicus and later Galileo, who espoused a he-

---

1) Co-President, International Physicians for the Prevention of Nuclear War

Einstein – Peace Now! Reiner Braun and David Krieger (Eds.)
Copyright © 2005 WILEY-VCH Verlag GmbH & Co. KGaA, Weinheim
ISBN 3-527-40604-2

liocentric universe in which the earth revolved around the sun and not the other way round. This undermined the ancient Aristotelian view of nature and promoted a new vision of Man's place in the cosmos. This view of nature, as an autonomous entity with its own laws, created a growing awareness of natural order, as opposed to divine intervention, and the first stirrings of the concept of *natural law* that later in the nineteenth century developed into the Darwinian view of the world.

The crystallization of the scientific method left in its wake a world transformed both intellectually and physically. Once thinking was liberated from the strait-jacket of dogmatic Christian theology, the old notion of a mysterious universe gave way to an understanding of a mechanical and orderly universe sustained by the law of physics.

Since the Renaissance and the Enlightenment, humanism is seen as a particular way of cultivating human intellect and expressing the belief that truth should be founded not on divine revelation, tradition or authority but on observation and reason. The aphorism of the Greek philosopher, Protagoras, became the motif of the Renaissance: *Man is the measure of all things.*

The history of the twentieth century, however, makes it difficult to view Man as the measure of all things, without a sense of self-mockery and cynicism. Today, we see Man as a much baser being, whose actions have despoiled the twentieth century: two World Wars, the holocaust, Hiroshima and Nagasaki, gulags, concentration camps and Guantanamo Bay, ethnic cleansing and genocide, environmental degradation and climate change. Asked to sum up the twentieth century, Yehudi Menuhin said, "It raised the greatest hopes, ever conceived by humanity, and destroyed all illusions and ideals."

As we begin the twenty-first century, we would not be far off the mark if we think of Man as weak, barbarous, inhuman, greedy, selfish, arrogant and self-destructive, perhaps never again as dignified and noble, or the measure of all things.

Nevertheless, the common optimistic view is that human beings occupy a unique place in nature because of their ability to reason and transform themselves and the world they inhabit. At the heart of humanism, therefore, there is a belief that humankind can achieve freedom and progress through its own efforts, affirming the writings of Karl Marx, based on his belief that human beings have the capacity to determine their own destiny. The question is: *for better or for worse?*

Human Conflict

Humankind has been prone to conflict throughout the ages. The written history of the world is largely a history of warfare, because the states in which we live came into existence largely through conquest, civil war or struggles for independence. The great statesmen of history have generally been men of violence for they understood the use of violence and did not shrink from using it to further their own political ambitions. Many of today's leaders are in the same mold.

The industrial revolution in the eighteenth century generated wealth and liberal ideas, which were largely channeled into military preparedness and military culture. Together with advances in science and weapons technology, wars in the twentieth century were waged with greater intensity and destructiveness, with the use of modern weaponry, such as the two atomic bombs that totally destroyed the Japanese cities of Hiroshima and Nagasaki in August 1945 and killed 200,000 people by the end of the year.

Thirty thousand nuclear weapons still menace the world today, as nuclear disarmament has reached a stalemate, despite the efforts of civil society, such as International Physicians for the Prevention of Nuclear War (IPPNW), which was founded in 1980, at the height of the Cold War, when a global nuclear war was a hideous possibility. In its ceaseless campaign to sensitize world opinion, create worldwide awareness of nuclear dangers, and persuade the nuclear weapon states to end nuclear proliferation and eliminate their nuclear arsenals, IPPNW has been inspired by the life of Albert Einstein.

Political Activism

Einstein's political activism began during the First World War. After the war, he made a speech at the Reichstag and warned that violence must not be fought with violence and that force breeds only bitterness, hatred and counter force. Such a warning would not be out of place in today's world where violence is endemic and militarism is on the rise.

In 1922, after the First World War, he spoke again at a meeting of the German Peace Federation at the Reichstag, calling for goodwill between peoples of different languages and cultures. Later, in a German pacifist publication, Einstein warned that war would undermine international cooperation, corrode culture, destroy intellectual freedom, bind the young to a culture of war, and cause economic depression. This is an uncanny echo of what is going on today in our contemporary world, where the protagonists of unilateral militarism scorn and undermine multilateralism and international cooperation, where mainstream media distort news and information to serve narrow national and corporate interests, where governments squander bloated military budgets and impoverish social services.

As a firm believer in world government and peaceful resolution of conflict, Einstein participated in the League of Nations, created in 1920 after the First World War with the purpose of achieving world peace. He resigned from the League Committee on Intellectual Cooperation in 1923, when France did not agree to arbitration over Germany's war-reparations payment, accusing the League of being a tool of the dominant nations. As the Covenant of the League was incorporated in the Treaty of Versailles and as the United States had not ratified the Treaty, the United States was excluded from the League. This weakened the League and it failed in the 1930s to deal effectively with the aggression of Japan in China, Italy in Ethiopia, and Germany in Europe, although it had organized international conferences, settled minor disputes and did useful humanitarian work. With the withdrawal of Germany from the League in 1933 and the rise of Adolf Hitler, it became clear that the League of Nations was impotent and incapable of realizing its goal of world peace. With war clouds hovering over Europe, Einstein reiterated the need for effective world government: "First, create the idea of super-sovereignty: men must be taught to think in world terms; every country will have to surrender a portion of its sovereignty through international cooperation. If we want to avoid war, we must try to make aggression impossible through the creation of an international tribunal having real authority."

In the current context of the United Nations, it is more than likely that Einstein would have railed against the unilateralism of dominant major powers that continues to weaken the United Nations. There are signs that the United Nations, which was created above all else "to save succeeding generations from the scourge of war," is also struggling to fulfill its objectives. The 2004 report of the Secretary-General's *High-level Panel on Threats, Challenges and Change* has recognized that the biggest security threats today go far beyond inter-state aggression and war. They extend to poverty, infectious diseases, environmental degradation, civil war, the spread and possible use of nuclear, chemical and biological weapons, terrorism and organized crime.

As a pacifist, Einstein believed that every person had the right to refuse military service or participate in war. He opposed compulsory military service and supported international protection for conscientious objectors. However, when Hitler came to power in 1933 and started building his Nazi war machine, Einstein modified his absolutist position on pacifism and instead supported the right of self-defense and the need for the democracies to arm themselves. He urged the United States to join the League of Nations and make it an effective instrument of international security, but by 1935 he pessimistically predicted that war would break out in two or three years.

Einstein was also concerned that science and technology were changing the nature and outcomes of warfare and that international conventions were ineffective in regulating the applications of science in warfare or in creating a strong system of international justice. He believed that militarism and preparations for war were economically and spiritually corrosive and damaging to societies. In 1930, he signed a manifesto for global disarmament, sponsored by the Women's International League for Peace and Freedom (WILPF), concluding that peace, freedom and security depended on disarmament.

In his search for enlightenment and new paths to peace, Einstein was open to new ideas and new visions. He corresponded with Sigmund Freud, Albert Schweitzer and Bertrand Russell, with whom he had discussions on the causes of war and remedies for war. Both Einstein and Freud agreed that an international legislative and judicial body was needed to resolve conflict and maintain security. In 1932, Einstein supported French Prime Minister Herriot's proposal for "a police force which would be subject to the authority of international organs."

As early as 1933, Einstein warned that powerful industrial interests, including the armaments industry, were attempting to sabotage efforts to resolve international disputes peacefully. He advocated greater understanding and awareness of the economic causes of war and identified the complicit role that the military-industrial complex played in the making of war and the greed for profits that betrayed our common humanity.

In 1937, Einstein declared that true pacifism reinforced international law, whereas neutrality and isolation, particularly by a great power, tended to contribute to international anarchy and consequently to war.

Nuclear Fission

One hundred years ago, in his *annus mirabilis,* Einstein published three landmark papers that proved the existence of the atom, demonstrated the validity of quantum physics and introduced the general theory of relativity. This laid the theoretical foundation for nuclear energy and ultimately the atomic bomb.

Although Einstein's groundbreaking formula, $E=mc^2$, theoretically indicated that nuclear fission in a very small amount of matter could convert matter into a tremendous amount of energy, he had discounted the likelihood of releasing energy from a molecule, saying, "It is something like shooting birds in the dark in a country where there are only a few birds."

When he realized that nuclear fission of uranium could set up a chain reaction and generate a large amount of energy that had the potential to produce an extraordinarily powerful explosion, he was persuaded by his friends and colleagues to write to President Roosevelt and warn him about the consequences should Nazi Germany acquire such nuclear technology. Unknown to Einstein, Roosevelt appointed an Advisory Committee on Uranium, which led to the fateful Manhattan Project and the subsequent development of nuclear weapons. This was the extent of Einstein's involvement in nuclear energy.

He was unaware that the United States had developed an atomic bomb until he heard the devastating news of the destruction of Hiroshima and the tremendous loss of life. This was a cruel irony for a man who was a dedicated pacifist and a passionate advocate of peace throughout his life. In November 1954, five months before his death, Einstein described his feelings about his fortuitous role in the creation of nuclear weapons: "I made one great mistake in my life....when I signed the letter to President Roosevelt recommending that atom bombs be made; but there was some justification – the danger that the Germans would make them."

After the Second World War, to assuage his conscience, he became more outspoken, believing that he had a responsibility to convey to the public the awesome reality of nuclear war and its threat to human survival. He was a leading figure in the World Government Movement and pleaded tirelessly for the abolition of all nuclear weapons through a supranational organization, empowered to maintain global security. Although the United Nations was a step in the right direction, Einstein believed that it was "a tragic illusion unless we are ready to take the further steps necessary to organize peace," such as abandoning national military power in favor of human security.

On 9th July 1955, Einstein and Bertrand Russell, together with nine other scientists, signed the Russell–Einstein Manifesto, which warned of the dangers of nuclear weapons and the continuing nuclear arms race that could result in universal death. They called on the US Congress, the scientists of the world and the general public to remember their humanity and subscribe to the following resolution:

> In view of the fact that in any future world war, nuclear weapons will certainly be employed, and that such weapons threaten the continued existence of mankind, we urge the governments of the world to realize, and to acknowledge publicly, that their purpose cannot be furthered by a world war, and we urge them, consequently, to find peaceful means for the settlement of all matters of dispute between them.[2]

New Nuclear Dangers

Einstein was persuaded that militarism and preparations for war frequently result in war and that nuclear weapons made war unacceptably risky. These are the realities that need to be carefully contemplated today when the only remaining nuclear-armed superpower espouses a new nuclear policy and speaks of justifiable unilateral pre-emptive military strikes against perceived threats to its security, including the use of nuclear weapons.

2) See: Appendix p. 287.

The new nuclear doctrine envisages a new triad of capabilities – nuclear and conventional offensive strikes combined with ballistic missile defenses and a new nuclear weapons production complex for designing, developing, manufacturing and testing new nuclear warheads, which are claimed to be 'usable.' In particular, the new doctrine expands the role of nuclear weapons from their Cold War function of deterrence to their use as legitimate tactical weapons on the battlefield. When seen in the context of the "war on terror" and the national security strategy of pre-emption and unilateralism, one has to conclude that any conventional war waged by the United States could escalate into a nuclear war.

The end of the Cold War has left the United States as the undisputed superpower and hegemon in an anarchic world of nation states, each preoccupied with its own narrow national interests. The security dilemma in international politics is the unpredictability of a state's intentions, which often leads to hegemonic alliances.

Hegemony, benign or malign, is distinct from an empire where the dominant, imperial state rules directly over subject territories, with the immediacy of force or threat of force. When a hegemon emerges, it does so from within a well-defined hierarchy among unequal nation states, maintained by the unequal distribution of resources and power. The hegemon leads because it is wealthier, technologically more advanced, and militarily more powerful than any other state. This is the position in which the United States finds itself today.

Despite their legal obligations under the Nuclear Non-Proliferation Treaty (NPT) to eliminate their nuclear arsenals, the nuclear weapon states refuse to comply and continue to retain nuclear weapons as the corner-stone of their security. More than that, the United States and Russia plan to develop new nuclear weapons and the United States, in addition, is intent on weaponizing outer space.

The unnerving realities of the atomic bombings of Hiroshima and Nagasaki appear to have faded from our collective memory, leaving a disquieting apathy about nuclear disarmament. To refresh our memory, nuclear weapons and their inhumanity must be seen for what they really are. Nuclear weapons are in a class of their own. They cannot be equated with chemical and biological weapons as weapons of mass destruction. Nuclear weapons are nothing less than illegal weapons of genocide that indiscriminately incinerate human beings in a matter of seconds. They are weapons of total annihilation.

Nuclear weapons must be eliminated and nuclear materials placed under strict international control to reduce the risks of nuclear proliferation and nuclear terrorism. The axiom of proliferation states that *the possession of nuclear weapons by any state is a constant stimulus to other states to acquire them.* First, it was the United States that possessed nuclear weapons. It was followed soon after by the Soviet Union, and then by Britain, France and China in quick succession. The nuclear club of five has now grown to eight to include Israel, India and Pakistan. North Korea may be another, and Iran is knocking on the door. The larger the club, the greater the danger of nuclear war.

There is a real danger that the United States is about to make the serious mistake of exchanging the potentially catastrophic nuclear dangers of the Cold War for the uncertain nuclear dangers of a unipolar world, challenged by international terrorism and nuclear proliferation.

Ethics

Today, science and the future of mankind are inextricably linked. Science has liberated human beings from mindless superstition, helped to offset the ravages of the elements and disease, and has become a major factor in the development of technology needed to sustain modern society. Thus, science has been legitimized, as we acknowledge the great good it can achieve in eliminating poverty and disease. But recently, its legitimacy has been questioned and scrutinized, as humankind confronts the grave dangers unleashed by modern technologies that threaten the continuation of life on the planet.

Some scientific advances in molecular biology and weapons technology suggest that science and technology need to be guided by universal moral principles. Much debate has revolved around whether problems are engendered by humankind's misuse of science or whether they are intrinsic to the scientific enterprise. Detractors of science argue that the very nature of scientific knowledge and its mode of enquiry are fatally flawed and that it is time to break free from the chains of stifling ideology and create an environment for transparency and accountability. Such an environment at the time of the Manhattan Project might have averted the tragedy of Hiroshima and Nagasaki.

As ethics are standards of conduct or social norms that define human behavior, scientists like any other profession must also be bound by an ethical code to ensure that their work contributes to the well-being of society. Scientists must re-examine the mythical concept, perpetuated by the scientific establishment, that science is neutral, purely objective and value-free.

There has been documented evidence of ethical misconduct in scientific research, as in the secret experiments on human beings during the Second World War and the Cold War. There is also increasing concern that the growing interdependence and interface between science and industry have generated ethical conflicts between scientific values and business values.

Despite these concerns, some scientists do take ethics very seriously. Scientists should emulate the ethos of physicians who honor the Hippocratic Principle: *First do no harm*. The Student Pugwash Group in the United States of America has initiated a timely pledge that has been signed by thousands of students worldwide:

> I promise to work for a better world, where science and technology are used in socially responsible ways. I will not use my education for any purpose intended to harm human beings or the environment. Throughout my career, I will consider the ethical implications of my work before I take action. While the demands placed upon me may be great, I sign this declaration because I recognize that individual responsibility is the first step on the path to peace.

These are the ethical values that shaped the life and work of Albert Einstein. He would be the first to acknowledge that science is fallible about its conclusions and ignorant about long-term consequences. In science as well as other spheres of human activity, we need a fundamental change in thinking, starting with the simple idea that it is wise to avoid unnecessary risk, especially if the consequences could be serious. This is the core concept of the 'precautionary principle,' which advocates that, in the face of scientific uncertainty, it is prudent to restrict or even prohibit an activity that may cause long-term or irreversible harm.

The precautionary principle was introduced as an ethical road sign and is now established in international declarations and agreements. The principle implies that responsibility for the safety of future generations and the environment should be weighed against the human needs of the present. In effect, the precautionary principle is a kind of insurance policy against our ignorance, our greed, even our arrogance, because we rarely understand or are aware of the risks until the damage has been done.

Paths to Peace

Global trends suggest that humankind has reached a dangerous crossroads. As members of the genus *Homo sapiens,* we have increased our knowledge at a phenomenal rate but not our wisdom. We are floundering in a sea of crises, not always aware that our world is crumbling and the light of our humanity is flickering.

Our globalized world is shaped by politics, economics and science, and corroded by deadly conflict, social injustice and ecological damage. While limited resources are squandered on military budgets, poverty, hunger and disease decimate populations. These are the silent weapons of mass destruction. Human security is as much about these ancient enemies of humankind, as it is about disarmament and the prevention of war.

The challenge is to redefine security in human terms and effect a paradigm shift from national security to human security. In other words, a shift from security of the state to security of the individual; from security through military power to security through human development; from territorial security to food, employment, health and environmental security.

There is also a leadership deficit. When many nations are led by leaders who mislead and misrepresent the truth, when rhetoric and double standards masquerade as diplomacy, when diplomacy itself is militarized, and when injustice divides the world, conflict and war ensue.

When conflict is allowed to fester and become chronic and intractable in the absence of political dialog, especially when one side is dominant and unyielding, the outcome is often political violence or terrorism.

Although one may feel pessimistic or apprehensive about the trends of our imperfect world, experience suggests that trends can be checked and altered by transforming attitudes, reinstating old values and strengthening institutions.

The state of our disorderly, polarized world strongly suggests the need for a new global ethical agenda for international politics and a renewed sense of global responsibility and global cohesion. This will entail a global consensus for binding values and ethical standards. The creation of a global ethic will depend upon what is established, not so much upon the norms accepted by states, as upon the norms embedded in institutions and practices that represent the wider consensus of universal values.

A global ethic is neither a global ideology nor a single unified global religion that transcends all existing religions. Nor is it a synthesis of all religions. Nor does it seek to replace the ethics of individual religions. A global ethic seeks to bring together what is already common to religions in terms of human conduct and moral values. Such a global ethic would be secular, universal and inclusive.

Despite pessimists and so-called realists, there are signs of a worldwide change in awareness about politics and governance, about economics and ecology, about disarmament and peaceful resolution of conflict, and about the partnership and interdependence between men and women.

National interests and narrow patriotism must give way to a higher allegiance to our common humanity and planet Earth. There is a South African term, *Ubuntu,* which refers to the spirit of community. *Ubuntu* means that we have a common humanity – each one of us is tied to the other and to the environment in a closely connected planet. This single most important concept of life needs to be understood and adopted universally by all peoples. This is the way we must live in peace so that we may survive.

Ron McCoy – IPPNW

Ron McCoy, born in 1930, is a Physician. For many years he has been the IPPNW Vice President. In 1985 IPPNW received the Peace Nobel Prize for their manifold Peace activities and campaigns to ban all nuclear weapons. Ron McCoy was President of the Malasian General Medical Council and in 1996/97 was a member of the Camberra Commission.

To What End ...? Changing Ethics?

Manfred Eigen

The background behind the choice of this title is that a slight shift in emphasis can easily result in a shift in meaning. Apocalyptic thoughts arise and they are thoroughly warranted in light of man's inability to use acquired knowledge sensibly.

Knowledge is not free of moral value.

Upon the death of Otto Hahn, the title of an article in "Die Zeit" read: "A discovery is neither good nor evil." The danger of such argumentation lies in that fact that it is so obviously correct to be trivial and one overlooks that every discovery – and I wouldn't know which one I could exclude – can have good or evil consequences. In my view it is equally as dangerous to believe that this problem can be solved by simply doing away with, prohibiting, or significantly limiting basic research. Instead, we must learn to use acquired knowledge *rationally*. Whether or not we bring a truth to light, it exists one way or another and will therefore be unveiled by someone someday.

A prime example, addressed in the article cited above, is the atomic bomb. A widespread opinion of our day was phrased in the following manner by Klaus M. Meyer-Abich.

> The atomic bomb was the first application of scientific knowledge in which there was no step between the so-called basic research and the technical development that could justify relieving the first phase of responsibility for the consequences that came in the wake of the second phase. The far-reaching consequences of scientific-technical development for the conditions of human life had been known since the beginnings of industrialization, but until this point, people believed that they could freely decide how basic research was to be applied." And he comes to the "inevitable conclusion": "The atom bomb was a direct result of basic research. *Thus, there is no basic research in the sense of a responsibility-free zone.* Rather, whoever contributed to the discovery of nuclear fission shares in the responsibility for the dead at Hiroshima and Nagasaki.

Einstein – Peace Now! Reiner Braun and David Krieger (Eds.)
Copyright © 2005 WILEY-VCH Verlag GmbH & Co. KGaA, Weinheim
ISBN 3-527-40604-2

I share with whomever concurs with this opinion the horror felt about the existence of these and *other weapons*. I hold the view – presumably as he does – that all efforts, all cunning, all knowledge must be used to destroy these weapons so that even their unintentional use can be prohibited. However, I also hold the conviction – probably contrary to many – that we have had no other choice than to live with this knowledge and to make *this insight* the basis of all of our further efforts.

Knowledge cannot be taken back, as Friedrich Dürrenmatt vividly illustrates to us in his parable about the physicists. Mankind has possessed, since the discovery of fire, *too much knowledge* to have ever been able to live in safety. On the other hand, we have *too little knowledge* – not only about our surrounding nature, but above all about ourselves – to prohibit the abuse of too much knowledge. Even if all atomic bombs were destroyed, their use would be but delayed in the event of war, never avoided. The danger of *the start* of war might, in some regards, even grow because the threshold would be lower. Yet the destruction of atomic weapons still makes sense because, while it wouldn't completely preclude the short-circuited deeds of an accidentally triggered atomic war, it would reduce the chances considerably. This sort of disarmament would have to take place on *all sides*. One-sided possession of any weapon leads with nearly inevitable certainty to their use – as Hiroshima and Nagasaki (and similar examples from World War I) prove. Treaties and their safeguards may not provide any guarantees, yet they are an initial step towards a possible solution. Such agreements do not, however, need be limited to nuclear weapons.

An imbalance on the side of knowledge is the greatest hurdle to concluding and upholding a treaty. Many see in knowledge the main cause for past and future catastrophes. I tend to see them in the gaps in knowledge. Limitations or moratoriums on *acquiring* knowledge, just as the prohibition of its possible *abuse*, would have to be contractually secured as well. If we could first guarantee that treaties are upheld, then they could limit themselves to preventing abuse. The acquisition of knowledge would also have to be inseparably coupled with the obligation to *disseminate* the knowledge acquired. This could not take place without contracts, and not without high moral standards on the part of those who are in a position to gain new knowledge. To prematurely and unilaterally forego knowledge – due to moral responsibility, for instance – harbors grave danger. It is convenient to lack knowledge when confronted with abuse.

In a world in which everyone abides by contracts, it hardly makes a difference if one has knowledge that can be used for war purposes, or not. In a world where no one abides by contracts, it would be highly dangerous if the moral side (whether one exists or not) had less knowledge and is thereby at the mercy of those who would abuse their knowledge. (This was Einstein's dilemma when he recommended that President Roosevelt develop the atomic bomb.)

First and foremost, it depends on whether we all adhere to contracts. Let us assume that this problem would not initially be resolved – the path does indeed appear long and arduous. Could we compensate this lack of adherence by constraining the acquisition of knowledge, perhaps by establishing a catalog enumerating what knowledge is allowed or even desirable?

If we were to consent to constraints on knowledge, then we effectively would no longer be able to conduct research today. Moreover, we would be forced to forget most of what we already know. Astrophysics would no longer exist because all knowledge of energy production by the stars would be extremely dangerous. The span of knowledge between Otto Hahn's experiment on uranium fission in 1938 and the ability to manufacture large quantities of material for a bomb (uranium-235 or plutonium) is roughly as wide as the difference in knowledge needed to make fireworks and the V2, both in terms of thought and technically. Einstein's recognition of the equivalence of energy and mass, Nernst's estimation of the difference in the mass of isotopes, and his inquiry into the "match" that could be used to ignite this "fuel", whose existence was already agreed upon in the 1920s, are all nothing other than the final confirmation of the correctness of such postulations. Otto Hahn published his results in the journal "Die Naturwissenschaften", making them accessible to the entire world, notably in a country that was preparing to attack its neighbors. He was not involved in any work whose aim was to develop a bomb. The path leading from Hahn's experiment to the construction of the bomb was in no way *short*. In Germany there were certainly enough scientists who "would have liked to" but "didn't have the courage". The effort involved in taking the idea and turning it into reality was in no way less than for any other scientific discovery. It was far greater when we consider the enormous facilities in

Oak Ridge required to split isotopes or the laboratories in Los Alamos and Livermore.

Is not what holds true for Otto Hahn's experiments also true for Hans Bethe's and Carl Friedrich von Weizsäcker's ideas about energy generation in the sun? These ideas could also initiate a series of events that could directly lead to the construction of Edward Teller's super bomb. Fortunately there has been no Hiroshima or Nagasaki for the hydrogen bomb, perhaps thanks to Edward Teller or Klaus Fuchs or others who ensured that knowledge was shared equally on all sides.

What remains that still may be researched? More importantly: *Who* is to decide what can be researched?

In the beginning, the researcher in his work is only sworn to the discovery of truth. The rule is: no end justifies the means. Neither methods nor findings are taboo. As in all questions of life, a careful choice between conflicting rights must be made. The scientist is not legally responsible for the object of his knowledge. However, he is responsible for *how* he achieved that knowledge. He must justify use of money and resources, and he must take care of what happens with his knowledge. In this case there *is* an obligation. In a democracy, whoever possesses special insight has special obligations. It is the obligation of the scientist to represent his insights and the necessities arising from them, regardless of whether they are politically opportune or popular. He must do this not because he is a scientist, but because he possesses knowledge that is not accessible to others, and his responsibility extends only to the realm of *his* knowledge. A politician, furthermore, must take into consideration whether his decisions will be carried by a majority.

Under the title "Changing Ethics", Thoma von Randow wrote: "Science has become of ill repute; even researchers must realise this by now and therefore shouldn't lack such instinct in light of the current public mood as to plead for animal experiments whose findings only satisfy our scientific curiosity." One could possibly argue about the last phrase as to what we understand by scientific curiosity, but von Randow leaves no doubt as to what he means when he concedes, "Certainly this statement is irrational and short sighted because the scientific findings of today can be our salvation of tomorrow. But what ethics have ever been rational?"

Ethics are nothing final and rationality is nothing preliminary, and neither of the two can simply be established by majority rule. In our country we have experienced worse majorities, although it is debatable whether every majority acting in public is indeed just that. Killing an animal requires in any case a decision in which a careful choice is made between conflicting rights. This is as true for the research laboratory as it is for the butcher. Respect for life cannot mean that we tolerate disease-causing agents and vermin in our living environment, though we must realise that we assume a highly biased stance by using the expression "vermin".

The question of the moral legitimization of scientific research holds a humanly biased component that is not to be overlooked. We cannot respond to it from a neutral position, from the position of a visitor from another planet, so to speak. Nor can we simply uncouple ourselves from our past. "He has a future who is in a position to say how the history that lies behind him can sensibly be continued through continuity or by breaking with the past. An awareness of our heritage is an elementary condition sine qua non for our future survival."

Manfred Eigen

Manfred Eigen, born in 1927, has been since 1964 the Director of the Max-Planck-Institute for Biophysical Chemistry in Göttingen, Germany. In 1967 he was awarded the Nobel Prize for Chemistry. His newest research considers self-organization of matter and the evolution of life.

The Role of Science and Technology in the Quest for a World at Peace[1)]

Jerome Karle

The past few hundred years of human history have witnessed remarkable developments in science and engineering. Man's profound scientific understanding and his engineering facility have expanded very rapidly. At the present time, it continues to accelerate,
affecting greatly almost all aspects of people's activities in all but the
least developed areas. Medical diagnostics and treatments, pharma-
ceuticals, appliances, transportation and communications facilities
are examples of the ongoing revolution that greatly affects individual
lives on a daily basis. The advances in health, comfort, convenience
and personal well-being are most evident.

Despite these inspiring accomplishments, the future is more and
more threatened with a deterioration of the quality of life and the proliferation of social inequities. It is therefore imperative that societies
make planning for the future a high-priority consideration. Appreciation of this need is increasing, having become much more widespread in recent years as people find the various manifestations of
the pressing problems part of their own life experiences. In looking
toward the future, it is incumbent on us to consider our past and
present circumstances in order to establish key issues, priorities and
necessary action. In this way, the impact of the threats may be minimized and cultural, ethical and humane values may not only prevail
but even be enhanced. Many individuals and organizations have already devoted much effort to these questions. Much still needs to be
done to bring societies and governments into effective action. It is
my intention here to present some impressions obtained from a
number of the studies that have been already made while emphasiz-

1) This talk was presented in the event series "Bridges ? Dialogues
Towards a Culture of Peace", hosted by Uwe Morawetz, Chairman of the International Peace Foundation in Thailand.

Einstein – Peace Now! Reiner Braun and David Krieger (Eds.)
Copyright © 2005 WILEY-VCH Verlag GmbH & Co. KGaA, Weinheim
ISBN 3-527-40604-2

ing the degree to which many of the current and potential problems are interconnected.

The objective is to identify major factors that contribute to or generate the conditions that threaten our planet. Such factors must be controlled so that their harmful effects may be minimized or reversed. Failure to do so could lead to great harm to the earth and life upon it. When we think of problem areas, we think of the environment, of human violence and warfare, of the many failings of human character, of the failings of leadership, of the interactions of economics, population and human psychology, of health the food supply, and we wonder about the extent to which many of these problems are interconnected and whether the interconnections can be sorted out so that there may be some hope of dealing with the problems effectively.

Environment

Much discussion has appeared in the literature of organizations dedicated to preserving a healthful and humane environment that concerns the many human activities that have severely degraded the environment. These activities threaten the habitability of major portions of the earth and perhaps all of it. How far do we have to look to find examples? Polluted air, waterways and oceans, ravaged lands, erosion, desertification, deforestation, and radioactive and chemical wastes are by now either common experience or common knowledge. Even the more esoteric threats that may not be as broadly appreciated or sensed, such as the depletion of the ozone layer and the accumulation of hothouse gases, have been brought to our attention. At which point does the damage become irreversible? How great an insult can the earth take and still bounce back? Have we passed the point of no return with some of these issues? Possibly, probably not. It would be prudent, nevertheless, to take immediate steps to minimize the effects of the various threats. This requires no less than world cooperation among societies as well as their governments. Can this be achieved?

There is another aspect of a proper environment that is important to consider, namely, quality of life. This means various things to various people, but it may be broadly described, as an aspect of life that goes beyond the minimal requirements of subsistence, shelter and basic health needs. For many of us, quality of life involves artistic, esthetic and intellectual components, human decency, high standards of behavior, peace, trust, kindly interactions among peoples with a respect for human dignity, and sufficient space in which to dwell. The amount of space for comfort may vary considerably among us, but I believe that most of us need room for privacy, thought and contemplation. This is not to be found readily in congested cities and congested living conditions brought on by high-density population.

High-density population impacts profoundly on the environment and all aspects of the quality of life. It is difficult to imagine an environmental problem that would not be improved by a decrease in population and made worse by an increase in population. Since the current trend is manifested by an explosion of population in many parts of the world, it is clear that this factor in environmental problems demands immediate attention.

The exceedingly high temperatures experienced in Europe this past summer may well have been a manifestation of global warming. If so, we have a clear example of one of the many serious consequences of ignoring the advice forthcoming from scientific and other technical groups.

Nature's Storehouse

There are many reasons for not destroying huge ecological areas such as the rain forests. A main one is that such areas are vast biological storehouses of substances having important potential to benefit humans. This derives from the fact that living organisms including both flora and fauna develop chemical mechanisms that protect them from hostile environments, predators, infectious agents and disease. Such problems have been solved in nature in a great variety of ways, most of which are yet to be discovered. The compounds involved can possibly serve humanity in many ways, as antibiotics, insecticides, herbicides, fungicides, preservatives, drugs such as anticancer agents and antimalarials, heart drugs and many other types of physiologically active substances. Destruction of ecological systems removes the opportunity to learn their secrets forever.

Economics and War

The earth is finite and the number of people that the earth can support is finite, under the best of circumstances. As a consequence, people have developed the concepts of a sustainable economy and a sustainable earth. Rarely is the expression sustainable economy heard. The term economic growth predominates. Certainly activities that replace others can grow and, to an extent, new activities can grow. Growth can also be associated with a rise of the average quality of life in the world. There are limits, however, and the limits arise from limits on natural resources, limits on need and demand, limits generated by environmental considerations, and constraints on total population. The latter statement provides a clear indication that future economic policy must be in harmony with the finiteness of resources and population, and the requirements for a wholesome environment.

Economic competition is often enhanced by limited resources and markets. In the extreme case, it leads to warfare. A major component of many wars has been economic competition. There have been, of course, additional issues, but it seems fair to say that economic motivation has predominated. Economic competition should not be expected to cease when warfare ceases. The experience of the post-World War II period, for example, is testimony to the fact that its underlying aspects of economic competition never disappeared, but simply has taken on another form. If and when the major players in industrialized states decide to confront economic reality by supporting and developing their human, humane and industrial capacities consistent with the broad concepts of sustainability and a large measure of self-sufficiency, they may be able to save their countries and future generations from severe economic hardship, if not destruction.

There is not much to say about war per se. It is enormously destructive, it dehumanizes and brutalizes survivors, damages the environment, wastes resources and may leave legacies of great potential harm to future generations. Nevertheless, warfare has been all too common a phenomenon in this past century. Societies must find other ways that are nonviolent to deal with their economic stresses and other perceived inequalities. Certainly, pouring arms into those societies that are most likely to use them offensively does not help matters. The large and apparently irresistible profit motive in the selling of arms makes control very difficult. The economic gain may be short range but the harm can be major and long lasting. There is not likely to be a net gain for any society in the long run.

There are other circumstances in which perceived short-term economic gains motivate inappropriate behavior and result in net economic loss and harm to society. Industrial pollution is an example. Pollution often occurs because industry makes short-range economic decisions, or because appropriate legal regulations are not in place or are not enforced. I have made a point of this to emphasize that economic and environmental problems require clear thought on the part of all concerned. The problems that threaten the world involve everyone. Solutions to the problems may require expert help but the implementation of the solutions involves us all. Societies and governments must be strongly involved and come together.

Gainful Employment in the Age of Automation

I recall reading an article several years ago that characterized a present economic dilemma, namely, that a new revolution is taking place in manufacturing, without being accompanied by a subsequent reduction in the work week. A main generator of this revolution is automation with its accompanying efficiencies. The author asserted that this is the first time a major manufacturing advance in efficiency was not followed by a reduction in the work week. The consequence of this circumstance is that advanced societies are becoming remarkably incapable of providing meaningful employment for a large fraction of its citizenry. Expanding populations can certainly add to the problem. In the present economic climate in many developed countries, the industrial response to monetary problems is to dismiss large numbers of employees and overwork others. This does not lead to economic stability for society as a whole. Manufacturers must also be prepared to anticipate rather than simply respond to dwindling supplies of raw materials and rapidly changing markets.

It would be most appropriate for societies to analyze the social and economic implications of the rapidly developing technologies, changing markets in the world economy, and other powerful forces that affect the quality of life in their communities. The problems are not insoluble. It would seem that successful societies would need to have a sustainable plan in which all the population would have the opportunity to make a contribution and thereby earn enough compensation to enjoy what most people would recognize as a decent standard of living. Can this be achieved in the atmosphere of competition that has characterized the industrial revolution up to this time? Is it possible to achieve sustainable economies and decent living standards without inhibiting the various aspects of individual initiative and creativity? There is a myriad of questions that societies may consider. The world, however, can not possibly benefit from continued procrastination in facing these issues or wait for the results of endless debate and deliberation. Too much debate and little testing is as bad for societies as it is for science.

What would need to be done, were governments willing, would be to take a careful look at resources, human and otherwise, set goals on the basis of high and presumably workable ideals and implement them. There would be the need for flexibility so that, in learning from experience, corrections could be made in a stepwise process. This would require governments to be composed of people, or advised by people, of the highest intelligence and personal standards and whose greatest satisfactions would derive from the good results that they could achieve. This is not as difficult to secure as may seem. There are many such people. The important step is for governments to be open to accepting well-motivated, well-thought-out plans and make difficult, self-less decisions, for the good of society unencumbered by parochial considerations.

A view of the history of the twentieth century suggests that the world is more than ready but ill-prepared for such high-minded developments. That certainly seems to be broadly true. Nevertheless, there are a few countries that have renounced war and appear to have pursued, for long periods, steps toward a humane society.

Ethics and Education

It is important that young people be trained from an early age to question contemporary concepts and ideas that swirl around them. They should be taught how to think and draw rational conclusions. Together with this, the development of high standards should come from the home as well as the school, if they do not come from the home, they must come from the school. The learning of purely factual material is, of course, indispensable, but education obtained in the absence of fine-tuned reasoning and an ethical sensitivity is probably a major contributor to current societal ills.

Valuable leadership qualities are the ability to apply sound intellectuality to problems and to use the results in an ethical fashion. It is to be expected that an educational system that emphasizes such qualities in general would raise the standards and the quality of life for society as a whole. In addition to learning how to think, a very worthwhile feature of an educational system is for it to be broadly based. Perhaps we may hope that, in the future, training in graduate institutions will be a prerequisite to appointment to high political office and that the entering students will be among the finest graduates of an educational system that promotes broad scholarship combined with ethical principles. The world can always benefit from great leaders with impeccable standards having broad intellectuality and a good sense of what may be achieved in practice.

Character

We are told by behavioral scholars that it is important to start to instill character into children at a very young age. Otherwise, in general, it becomes much more difficult. This appears to be a subject that merits careful consideration, since it seems to be vital for the achievement of more humane and ethical societies in the future. In addition to building character, it is important to teach young people to have inquiring minds and well-ordered priorities. They must learn to think, gather information and draw rational conclusions based on the information.

Considering the threats to future existence, it is self-defeating to persist in maintaining many current societal priorities and values. There needs to be, of course, flexibility and respect for a variety of opinions, but when ethical standards, for example, are quite low, it should be possible to decide on which standards should be raised and in what way. We make laws to try to correct or prevent unacceptable behavior. It is much better when people's ethics, humane principles and self-respect make the laws unnecessary or, at least, seldom needed.

Summary Remarks

Interconnections
There are a number of issues concerning the interactions of humans with the earth and with each other. Some of the more important ones are listed in alphabetical order below:

Economic matters
Education
Environment
Ethical, humane and wholesome behavior
Finiteness of resources
Health care
Illegal drugs
Population
Priorities and values
Quality of life
Sustainability with limited resources
Transportation

These issues are far from independent of each other. They are, rather, grossly interdependent. This interdependence leads to the concept of "indispensables", namely, those issues that are particularly indispensable to the achievement of widespread improvements in the human condition. In fact, without proper handling of the indispensables, the attempt to achieve improvements in a broad sense by appropriate treatment of the other issues has little chance of success.

Indispensables
Population control is an indispensable. I am in agreement with many others who have indicated that unless population is brought under control, it will not be possible to enhance the human condition on a broad scale. Any progress will be eliminated by the huge numbers of people who must be supported if the unbridled increase in overall population persists. Population control must be achieved.

A second indispensable is sustainability. The control of sustainability must be brought into the handling of environmental problems. Activities that degrade the environment can no longer be tolerated. Renewable energy sources, for example, must replace the burning of fossil fuels. Economic decisions impact greatly on the sustainability of the environment and, in fact, on the viability and sustainability of an economy. It was pointed out in a talk that I had heard some time ago that the inhabitants of Easter Island, located off the west coast of South America, used up all the resources of the island to the extent that they did not have enough food to subsist and did not have enough wood to build boats on which to escape from the island.

Proper human behavior is an indispensable. The amount of animosity of various groups toward their neighbors in many areas of the world is a serious impediment to progress. The ease with which seemingly cultured societies can be transformed to behave in a barbaric fashion is another sad lesson of the twentieth century. Such threats of unbridled hostility and violence must be curtailed if a proper environment for peace and stability is to be achieved.

Barriers

There are various barriers that can interfere with the broad attainment of a quality existence. The problems that need to be overcome are formidable. They are manifold and extensive. Great leadership in the world is required to overcome societal indolence, selfish motives, lack of character, perverse priorities and values, widespread unwholesome life styles, lack of cooperativity, extreme poverty, educational limitations, violence and the lack of suitable mechanisms to settle disputes peacefully, exploitation of the earth and of people and a population explosion that is out of control.

Societies whose survival is marginal can not readily give much attention to the issues that are raised in this presentation. It is very difficult for societies or individuals on the edge of survival to change the patterns of their lives. Many marginal societies are overpopulated, are under severe economic pressures, and have low levels of heath care.

Concluding Question

We live in a world in which there are major inequities within societies and among societies. It is also a world whose future is greatly threatened. Do we want nature to take its course, a path that is often extremely harsh, or will the world's population and leadership be willing and able to take the major courageous steps required to mitigate nature's harshness, preserve the earth and generate a more equitable and humane future?

The role of scientists and other technical people has, at least in part, been observed for many years. This concerns the many warnings that have been made public concerning such matters, for example, as global warming, pollution, the loss of important information associated with the destruction of ecological areas such as rain forests and the bad effects of fluorine hydrocarbons on the ozone layers in the atmosphere. Scientists and engineers can provide education and modern equipment to countries that could benefit from such attention. This is starting. In the future, it is expected that education, health and technical assistance will increase.

For those matters that require more attention than that which would come from technologists, it is likely that organizations of scientists and engineers would be interested in considering participating with other groups to attain worthwhile goals. An example would be biennial meetings of scientists, engineers, members of peace organizations, world leaders, representatives of the World Health Organization and officials in the United Nations.

Jerome Karle

Jerome Karle was born in 1918. He holds a M.A. degree in Biology and PhD in Physical Chemistry and. His research has been concerned with different theories and their application to the determination of atomic arrangements of substances in various states of aggregation: gaseous, liquid, amorphous, solid, fibrous and crystalline. In 1985 his work in crystal structure analysis was recognized by the Nobel Prize in Chemistry. Karle is a member of the Committee on Human Rights of the National Academy of Sciences and Advisor to ChildRight Worldwide.

Science and Society – Some Reflections

Jean Maria Lehn

Science provides knowledge that technology transforms into means of action. Their impact on society will depend on the effective use that is being made of them. Chemical and physical sciences have profoundly modified the living conditions of mankind, in particular chemistry has played a central role. Its creative power has made available a range of new materials and processes for the transformation of matter. These have often been perceived as unnatural and opposed to the natural ones. As a result the question of a control by society has become more and more actual. This requires information of the public and of decision-makers for evaluating the potential impact of technology on society as well as for preserving the necessary freedom of scientific research and of quest of knowledge.

Science offers most exciting *perspectives for the future generations.* It promises a more and more complete understanding of the universe, an always greater creative power of chemical sciences over the structure and transformations of the inanimate as well as of the living world, an increasing ability to take control over disease, aging and even over the evolution of the human species, a deeper penetration into the working of the brain, the nature of consciousness and the origin of thought.

Science and its implementation through technology have transformed and will continue to *transform society* in many ways. For instance, through the extraordinary development of electronics and new materials, physical and chemical sciences have made possible the advent of the age of information and communication, abolishing time and distance, bringing people together through fast and secure transport. Major effects on work habits and employment have resulted. Thus with phone, fax and electronic mail, will there be post-offices and post employees in the future? What should one and can one do to conserve social contact when much of the work will be

Einstein – Peace Now! Reiner Braun and David Krieger (Eds.)
Copyright © 2005 WILEY-VCH Verlag GmbH & Co. KGaA, Weinheim
ISBN 3-527-40604-2

done by remote means, at home and long-distance? Progress in basic knowledge and the development of new technologies lead from labor intensive to intellectually intensive activities. This causes a recomposition of the employment scene. Many jobs disappear but many new ones are created. All areas of human life and endeavor, for the individual as well as for the society, are concerned. Discovery requires researchers, development needs engineers and technicians, applications are in the hand of workers. All must be assisted by management, secretarial and maintenance staff.

Novel technologies also induce a shift from heavy industries, making strong demands on raw materials and energy, to much more environmental-friendly activities.

Many new jobs and activities are being created, which have as a main feature that they are of higher intellectual content than the hard labor jobs that are phased out.

Biological sciences and technologies are providing entirely new perspectives to our understanding of and action on the living world, for health care, food production, environmental control. They will have strong impacts on social and personal relationships, family structure, law and ethical values.

The resolution of pressing *economic and social problems* cannot be envisaged without fundamental discoveries to open the way to new technologies. It is thus a two-fold challenge, intellectual and technical. To meet it becomes ever more fascinating. One must have the courage to match the dangers, the ambition to meet the challenges, whether they originate from the abyss of our ignorance or from the spectre of our crises. Such perspectives require science and research. But science and research for what?

Firstly, research is needed for the *acquisition of new knowledge.*

One could imagine that basic research would be an entirely intellectual activity, a human endeavor like art, with no link to any application. One might consider that present basic knowledge is sufficient for all applications and technologies one would like or need to develop. This would not only be short-sighted and suicidal, but would also stop all progress. It would renounce finding higher level solutions to present-day problems.

In a time where questions about the justification of continued scientific research are being asked, we must take a strong stand. Between continuing our investigations and stopping them, there is only one valid option, we must continue, because it is the fate of mankind to pursue the quest for knowledge, because it is the only way to solve problems that go unsolved, because we cannot, we do not have the right to close the road to the future. The generations to come would not forgive us if we decided that the level *we* have reached is sufficient. Also, the option to stop is a totally egoistic one; only those who already have plenty can ask themselves such a question, not the others. For those who do not benefit at present from the progress made, we must continue, with a strong sense of commitment and of responsibility.

Secondly, progress in research is needed for the *development of novel technologies.*

Long-term basic research is essential for the progress of our ability to shape the world around us, to free us from the chains of evolution and to open our window to the universe and to the future. Putting knowledge into practice has provided and will continue to provide novel and increasingly powerful technologies giving us new freedoms, new ways of life and new means of action. We must seize the chances they offer.

The development of the *environmental protection* movement, especially in its more extreme form of ecological fundamentalism or deep-ecology, due at least in part to sloppy operation of industrial plants and to disagreeable (not necessary harmful) pollutions, has led to the exacerbation of *the natural versus unnatural debate.* One must at the outset distinguish between the dangerous and the disagreable (not to speak of the plainly wrong). If quality of life benefits from the containment and elimination of both, only the first is threatening and it can usually be controlled by means of existing technologies; but one has to apply them and pay for it.

When irrationally implemented and shortsighted, environmental protection may have very perverse effects. I may cite a particularly tragic illustration taken from *A Review of the Greatest Unfounded Health Scares of Recent Times* published by the American Council on Science and Health. It concerns the pesticide DDT and malaria. "The ban on DDT was considered the first major victory for the environmentalist movement in the US. The effect of the ban in other nations was less salutary, however. In Ceylon (now Sri Lanka) DDT spraying had reduced malaria cases from 2.8 million in 1948 to 17 in 1963. After spraying was stopped in 1964, malaria cases began to rise again and reached 2.5 million in 1969. The same pattern was repeated in many other tropical – and usually impoverished – regions of the world".

DDT spraying in Africa began in the 1950s and greatly reduced the incidence of malaria. The ban on the use of DDT in Africa has led to a disastrous re-emergence of malaria, which now kills 3000 African children a day.

In Zanzibar the prevalence of malaria among the populace dropped from 70 per cent in 1958 to 5 per cent in 1964. By 1984 it was back up to between 50 and 60 per cent. The chief malaria expert for the US Agency for International Development said that malaria would have been 98 per cent eradicated had DDT continued to be used.

In addition, from 1960 to 1974 the WHO screened about 2000 compounds for use as antimalarial insecticides. Only 30 were judged promising enough to warrant field trials. WHO found that none of those compounds had the persistence of DDT or was as safe as DDT. (Insecticides such as malathion and carbaryl, which are much more toxic that DDT, were used instead.) And – a very important factor for malaria control in less-developed countries – all of the substitutes were considerably more expensive that DDT.

Taking a stand with respect to irrational and emotional approaches, many of us scientists have endorsed the *Heidelberg Appeal* addressed in April 1992 to the heads of states and governments attending the Rio de Janeiro Conference on Environment and Development held on June 3–14, 1992.

A major issue for all is that of *conservation* of resources, raw materials, energy. It requires us to give up careless and wasteful behavior in a throwaway society. This is a question of individual and collective attitude and not a consequence of science and technology. Science provides the knowledge and technology the means of action. The decision about how to make use of them is ours, as individuals and as societies. This is of course directly linked to the problem of world population, which, however, is not just determined by the number of people alive on the planet. It also depends on how these people behave.

The pursuit of a *sustainable development* for all regions of the world requires that we seek to achieve minimal waste, ideally zero waste, of resources, whenever the global balance of materials and energy makes it feasible.

This means maximum use of all components of present-day natural resources as well as the use of science and technology for the development of highly efficient processes and of new products, such as for instance novel types of engineered plants, so as to minimize energy consumption and waste generation.

The suspicion against science and its technological realizations are too often carried by a diffuse fear: there are things that one should not touch, for fear of catastrophe. This myth of an intrinsically "pure" Nature whose harmony would be disturbed by man is underlined by a quasi-religion. But nature is totally indifferent to man, it is not good and not bad, it just *is*. It is mankind that has to shape the environment in which it feels good to live, which may of course differ widely with climate, culture, etc. Nature also is not constant, and the idea of conservation is in itself anti-natural; the biosphere has been and is being modified through natural causes; the environment changes constantly and has always done so. The question is what change and at what rate; an answer is to call for a responsible management of the biosphere, and of the resources of the planet.

The *technologies of life*, resulting from the extraordinary progress made in understanding life processes and the ability to act upon them, appear to tamper with a basic mystery and to lift an interdict with the risk of unleashing uncontrollable forces. It was the case already with atomic energy, but to some extent electricity and automotive power may have evoked similar fears in older times. We have only moderately well controlled atomic energy, but it could have been much worse. Of course, the story is not finished, but this first and painful examination has perhaps prepared us to bring somewhat better under control the new powers that humanity gives itself through science and technology, and in particular with the technologies of energy and the biotechnologies. The realizations and potentialities of genetic engineering have aroused many reservations, in some countries more than in others. But the benefits that they can bring are countless, in agriculture and food production for instance, but above all to human health. Substances extracted from natural sources may be contaminated by compounds that present risks to health. Biotechnologies may permit us to circumvent the problem. Thus, synthetic vaccines may be safer that natural ones. The production of human growth hormone by genetic engineering gives a product devoid of the prion that infects the same substance of natural origin and causes Creutzfeld–Jakob disease. Numerous other such cases can be found, a particularly vital one being that of the production of factor VIII for blood transfusion without risk of infection by HIV.

In the historical perspective, natural evolution has led to a being – man – that has progressively become able to take charge of itself. And, except for a natural or provoked catastrophe, mankind will inevitably end up controlling its own evolution. Let's be clear, if it acquires this capacity it will make use of it, sooner or later, for better or worse. It is certainly still far away, but we must take this eventuality into account. And after all, it will only be natural, if man, the product of natural evolution becomes able to replace random evolution by controlled evolution. One may say that *man modifying man is contained in man*.

The respect of life has led to the axiom of the "sacredness" of life. But life does not respect itself; living species feed on each other. If life is to be called sacred, then it is not by an external reference, but because *we*, human beings, are living organisms. Since not every form of life is equally sacred, a provisory 'axiom' may be to respect and preserve those features that allowed the formulation of and the quest for that very axiom itself: consciousness and thought. That life has eventually led to consciousness and thought is the best ground for a general respect of life itself. The basic value can only be a metabiological one, resting on something that transcends life even if it necessarily is a product of life. Life has evolved a species that is able to question its own making and to transcend what has given rise to it.

The French writer Vercors has defined humans as "*animaux dénaturés*" as "denatured animals". A being that is separated from, ripped out of nature, living in it but able to observe it, question it, investigate it from a distance, from the outside, conscious of its own separate existence. Such a distinction has led to the development of these powerful means of scrutiny and discovery, of invention, of creation and of action, that are scientific research and technologies. It resulted in the *natural/unnatural dualism* that characterizes the ambivalent relationship of man with nature, and in particular to the impassionate discourses, the heated debates and the extreme attitudes that cloud many important issues requiring a cool-headed, rational approach.

My own field, *chemistry* suffers from suspicion and rejection on the part of our societies. Here also there is a big misunderstanding. One opposes "chemistry" and "nature". But, a compound is always chemical, be it natural or not, be it generated by a plant or an animal, or produced in the laboratory. A natural substance has no reason to be less toxic than a synthetic one, which is in fact usually in a purer state. In recent years, Bruce Ames has applied to compounds found in common foods the widely used test that he had developed for determining the carcinogenicity of artificial chemicals. The conclusion was quite startling: "Dietary pesticides (99.99%) all natural"!

Let me at this stage, briefly evoke chemistry and its *role in science and society*.

Indeed, chemistry plays a *central role* both by its place in the natural sciences and in knowledge, and by its economic importance and omnipresence in our everyday lives. Being present everywhere, it tends to be forgotten and to go unnoticed. It does not advertise itself but, without it, those achievements we consider spectacular would not see the light of day: therapeutic exploits, feats in space, marvels of technology, and so forth. It contributes to meeting humanity's needs in food and medication, in clothing and shelter, in energy and raw materials, in transport and communications. It supplies materials for physics and industry, models and substrates for biology and pharmacology, properties and processes for science and technology.

Chemistry has traced its path in the history of the universe. In the beginning was the *Big Bang*, and physics reigned. Then chemistry came along at milder temperatures; particles formed into atoms; these united to give more and more complex molecules, which in turn formed aggregates and membranes, defining primitive cells out of which life emerged.

Chemistry is the *science of matter and of its transformations*, and *life* is its highest expression. It provides structures endowed with properties and develops processes for the synthesis of structures. It plays a primordial role in our understanding of material phenomena, in our capability to act upon them, to modify them, to control them and to invent new expressions of them.

Chemistry is also a science of *transfers*, a *communication center* and a *relay* between the simple and the complex, between the laws of physics and the rules of life, between the basic and the applied. If it is thus defined in its interdisciplinary relationships, it is also defined in itself, by its object and its method.

In its *method*, chemistry is a science of interactions, of transformations and of models. In its *object*, the molecule and the material, chemistry expresses its creative faculty. Chemical synthesis has the power to produce new molecules and new materials with new properties. New indeed, because they did not exist before being created by the recomposition of atomic arrangements into novel and infinitely varied combinations and structures.

By the plasticity of the shapes and functions of the molecule and the material, by its creative power as well as its role as a relay, chemistry is not without *analogy to art*, a *process of transfer by the created work*.

Chemistry has been evolving over the years towards an increase in *complexity* and in *diversity*, from molecules to materials, from structures to architectures, from properties to functions. Thus, beyond the molecule one witnesses the emergence of a *supramolecular chemistry*, my own field of research, the chemistry of molecular assemblies, of self-organizing systems and of complex functions. In addition to the exploration of the interface with biology, a definite emphasis lies in non-natural species, possessing a desired chemical or physical property. It opens wide the door to the creative imagination of the chemist at the meeting point of chemistry with biology and physics.

Thus, chemistry is a science and an art. It is also an *industry*: each scientific component of the discipline has its industrial counterpart. It has for this reason a very marked impact on economic and social life.

It is not surprising that chemistry is called upon more and more often to face a number of new, or increasingly important, socio-economic imperatives associated with geopolitical phenomena. Some result from new economic conditions in the industry – costs and availability of raw materials and energy – others from the reorientation of the chemical industry (the necessity to produce products with high added value, possessing new properties), and finally still others from social preoccupations concerning the environment and the quality of life, such as improvements in working conditions, the safe use of products, the protection of the community, the fight against pollution. Each aspect of human activity therefore depends upon a better knowledge of chemistry and on its progress, and can be improved by it.

This being said, it is clear that the chemical industry must avoid degrading the natural environment whether for reasons of toxicity, simple unpleasantness or even on esthetic grounds. A plant can perfectly well function without emitting disagreable fumes or odors if one wants to do so and is prepared to pay for. If it is imperative that industrial activity be both not dangerous and not disagreable, it is unfortunately not always possible to accurately assess the risks. These deficiencies are a further incentive for pursuing research, but more knowledge does not solve everything. Between a clear danger and a possible risk the choice is easy. Between pleasantness and risk, it is much less so. Thus, the warnings about the dangers of smoking do not prevent people from smoking. We all have to take decisions and must be ready to bear the consequences.

Zero risk does not exist. *Risk* appears with life. Zero risk is a dead world. The desire to systematically eliminate all risk may also become a threat to freedom and democracy. It may lead to the elaboration of vast sets of regulations, justified or not, but usually anyway insufficient and that may hinder our freedom of action. At a certain stage, regulations become an unacceptable limitation to freedom. Of course, the answer is not to go back to the past, but one should not forget that Pasteur, the centenary of whose death was celebrated worldwide in 1995, experimented his vaccine against rabies in conditions that would make one shudder today. Similarly, it has become commonplace to say that such a fabulous drug as aspirin, which is working wonders one continues to discover, would probably not pass the regulations in vigor today. As for Pasteur's vaccine, it was a gamble that in present days appears quite dangerous but which placed in the context of the time was justified. One may take all possible precautions, every decision implies taking a risk. Later, in the light of new knowledge acquired, some decisions may be seen as a tragic error, but history cannot be rewritten and the mere fact of living is taking a risk!

Our duty is to *optimize the chances and minimize the risks*. A glass that is half-full to some people, is half-empty to others; but the full half, one can drink, whereas with the empty half one cannot do much (except try to fill it)! The much-discussed precautionary principle is a dead-end. And doing nothing can be a much greater risk than doing something!

An attitude that some adopt is to say "Let's not pass on to our descendants the consequences of our errors". There is evidently a very judicious side to this, provided it does not induce a precautionary attitude that would lead to stopping research and thus deprive future generations of knowledge that may be very useful to them. Our descendants will continue to evolve intellectually, culturally, materially. They may with hindsight, adopt points of view quite different from ours. To stop the machine would deprive them of the possibility of further development and would prevent them from succeeding where we failed. *This* is our responsibility, we have no right to force them into such a situation and to hand down judgments in their place. They may be wiser than we are.

Despite all these reservations, environmentalists have played and do play an important role. They have brought forward certain drifts that may become dangerous; they have led people to reflect and not to be carried away by the fascination of progress at all costs. Would this awareness have been gained without them? Perhaps, but probably more slowly and less clearly.

Confronted with societal mutation, the scientist is not a citizen like any other. Not that he or she has other rights, but additional duties, *responsibilities* and a better knowledge of the situation. Nothing is black or white, especially at the start of a study. But the responsibility of the scientist is to make others aware of the potential harmful consequences of his/her work, of the dangers of the instrument he/she is developing, to the best of his/her knowledge. The responsibility and eventual culpability lie in the words "to the best of his/her knowledge". How can one judge? Who will do it? At what stage? Rewriting history is not acceptable.

The discovery, the invention, the machine or the device are one thing, the use one makes of them is altogether another one. The technological potential of a discovery is assessed on the basis of criteria that often have nothing scientific within them. Political, strategic, economic choices are being made. In the area of health, particularly, launching new programmes may involve agonizing decisions at each level. Financial resources being necessarily limited, what fraction should be devoted to heavy techniques such as organ transplantations that carry great hopes but are very costly? Should one not systematically favor areas that would be profitable to the larger population? A decision taken today may be questioned tomorrow with the appearance of new data. Thus, the build-up in micro-organisms of resistance to antibiotics that leads for instance to the strong re-emergence of tuberculosis, will require new investments in research on the development of new antibiotics. What fraction of the available funds should be allocated to this versus antivirals, for instance?

In such choices scientists in general (and physicians in particular), must play a determining role in the decision taking. The same holds for environmental questions.

Only those who are competent in a given field will be able to advise and should advise; they have to have the confidence of the others, knowing, however, that they are nevertheless fallible. To use a metaphor: would you wish that other people than the competent ones choose the pilot of the plane you are taking? Should all the people in the cabin participate in the choice?

But if some must decide, the rules have to be clearly defined at the outset and the responsibilities delineated beforehand. If not there may come a time where nobody will be willing to advice or to decide on sensitive matters.

Broadening the scope, which should be or can be the *participation of the citizen* in the choices that affect the society and his or her life directly? Is direct democracy the answer or parliamentary democracy, where supposedly(!) competent, trustworthy and reliable people are selected to handle the affairs of the city and the state? Direct democracy may be an ideal but is it realistic? It first of all requires education of everybody to a level that would permit them to judge and decide on the issues concerned. Education implies providing information plus judgment, thus giving the ability to evaluate. The development of the means of information and communication could in principle make direct democracy possible, but is it feasible and when? Progress in communication may have the opposite effect of what one may hope: it may give power to those who do not know, to deception, rather than bring knowledge. Democracy versus demagogy, education versus deception, confidence versus suspicion, competence versus ignorance, these are the alternatives.

Competence, confidence, reliability are major requirements for a proper functioning of our democracies, which are therefore in fact "gnoseocracies" or "noocracies", government, decisions taken by those who know, or at least should know!

A crucial question concerns the so-called *developing countries*. Will the gap that separates them from developed countries get wider and wider? If East/West relations have dramatically changed for the better, a continuing and aggravating problem into the next century is the unacceptable *North/South imbalance* and resulting strains. It is the responsibility of the developed countries to offer solutions.

The 1992 Rio Conference put forward the concept of "sustainable development". Is it inescapable and acceptable that certain fractions of mankind develop at the expense of the others? Could progress have been more evenly distributed over the planet? It was probably inevitable that some countries, some regions of the world develop faster than others and that a parallel evolution of all would not have been feasible. But, then, it is the duty of those who had that luck, to redistribute knowledge and wealth. The responsibility of those who have enjoyed these benefits, scientists and citizens of developed countries, lies, beyond sterile expressions of culpability, in contributing to the goal that the increase in knowledge serve to also improve the life of the rest of mankind. It is for this reason too that it would be criminal to stop the course of scientific and technological progress.

It is perhaps not totally utopic to hope that the accumulated knowledge and wealth and the very efficient advanced technologies resulting from research in the developed countries, might provide means for the rapid progress of the less-advanced ones, by helping them to jump over intermediate stages of development, such as those of the industrial age and of heavy industry, directly into a "high-tech" era based on highly advanced technologies that are more economical and much less demanding in raw materials, resources and energy. The development realized in some countries may then serve for the good of all.

Such direct entry into an era of soft, human-friendly and environmentally conscious technologies, would amount to a sort of developmental short-cut, a *development-shunt*.

For instance, a country that has an unsatisfactory telephone system will not have to lay more wires but can directly go to cellular phones, and may thus even be at an advantage over countries having a developed classical phone network.

Technology transfer should thus be effected at high level; it implies to introduce photovoltaic or nuclear electricity generators not coal-fueled power stations, high-performance materials not steel mills, etc. Such high-tech transfers related to energy, materials and health imply that *we* do the research that these countries cannot do at present (for instance on tropical diseases) but also that we set the course that will allow them in the future to take full advantage of science and technologies. And this requires education and the transfer of knowledge.

Science education in our schools, colleges and universities as well as for the general public must be a major priority, so as to train the researchers and discoverers of tomorrow – to lift irrational fears and rejections – to develop the scientific spirit, the scientific attitude, in order to fight the obscure, the deceitful, the irrational.

Beyond the general progress of knowledge and the technological development, the most important impact science can and must have on society is the *spirit* that it implies, the scientific, rational approaches towards the world, life and society.

Education, science and technology may collide with tradition and hurt beliefs or social structure. We must be prepared for that and take it into account so as to overcome it. The installation of a solar-powered water pump accessible to everybody in a village of a developing country may destroy a traditional structure where power was in the hands of those who controlled the water supply. As another example, a medical study concluded that sexual relationships for a year or more before fertilization markedly decreases some health risks (eclampsy and hypertension) linked to pregnancy, by making the woman immunologically prepared to conceive through contact with the foreign male spermatozoa cells. Thus, sexual cohabitation without conceiving appears advisable. This goes squarely against some religious convictions and bans. Not to speak about methods of fertility control (such as RU 486), to which some circles, where fundamentalists from all sides are hand in hand, manifest strong opposition. Thus, science offers new freedoms but mankind has to learn to live with them.

Helping to *bring science to everybody* is also the responsibility of the media, which at present play at best a minimal role, often a deplorable one. Major improvements must be brought about, and a broad public would certainly be much more interested in high-quality science programmes than many media bodies like to believe. There is here an extraordinary contribution to be made at the interface of science and the society.

A major factor is the emergence and development of *information and communication networks* and the resulting modification in the dissemination and transmission of knowledge. Humanity has transferred knowledge and culture through education. The worldwide accessibility of information may lead to profound changes. The whole educational system will have to be adapted; in addition to basic training from nursery school to university, more and more importance will have to be given to „continuing education", lifelong regular updating so as to provide the flexibility required for coping with rapid changes. Education by parents and teachers will be more and more complemented by self-education, children educating themselves through direct access and individual choice. The role of the teacher will change profoundly; at the limit, the teacher as we know him/her could disappear. The evolution may well be towards a *global socratic maieutic.*

The development of communication technologies could also bring about the end of the interpreter through the advent of instantaneous automatic interpretation and translation; as a major corollary it would preserve the diversity of the languages and cultures of the world by allowing direct communication in one's own language, in a sort of fully transparent Babel tower, where everybody would understand everybody and have access to all cultures! Information networks and communication are neutral; the questions are how they are used, what they make possible and what should be allowed, thus raising ethical considerations. But restrictive rigid rules should not be enacted; it is sufficient to progressively outline regulations that may be adapted as the activities develop.

A very actual issue concerns the situation of the scientist with respect to *ethics and society.* It is my strong opinion that the scientist has first of all *general responsibility to the truth* and only then is there responsibility to the society and the world at the particular time in history. Ethics is a function of time, location and knowledge. Pursuit of knowledge and truth supersedes present considerations on what nature, life or the world are or should be, for our own vision can only be a narrow one. Ethical evaluation and rules of justice have changed and will change over time and have to adapt. Law is made for man, not man for law. If it does not fit anymore, change it.

Society will react to changes introduced by science and technology like a large organism; it will evolve and adapt under the pressure of new ways and means in a sort of *societal Darwinism*.

Some think that it is being arrogant to try to modify nature; arrogance is to claim that *we* are perfect as we are! With all the caution that must be exercized and despite the risks that will be encountered, carefully pondering each step, mankind must and will continue along its path, for *we have no right to switch off the lights of the future*.

These perspectives for the future of science, for *our* future, have already been expressed in most fitting terms by this quintessence of the artist-scientist, *Leonardo da Vinci* when he wrote:

"Where nature finishes to produce its own species, man begins, using natural things, with the help of this nature, to create an infinity of species".

Prometheus has conquered the fire and we cannot give it back. We have to walk the way *from the tree of knowledge to the control of destiny*.

Jean Maria Lehn

Jean Marie Lehn is a Chemist and was born in 1939. A major research effort is presently devoted to supramolecular self-organization, the design and properties of "programmed" supramolecular systems. In 1987 he was awarded the Nobel Prize for Chemistry.

Technology, Tolerance and Terror

John C. Polanyi

The centenary of Albert Einstein's dazzling emergence on the world stage in 1905 is also memorable as the fiftieth anniversary of the ending of the second of the two World Wars Einstein was to experience.

On February 11th of 1945 three world leaders met in Yalta to give shape to post-World War II Europe. Stalin was the host. He jocularly introduced the head of his secret police, the mass-murderer Lavrenti Beria, to Roosevelt and Churchill as "Russia's own Himmler" (Himmler being the Nazi's mass-murderer). In the decades between 1905 and 1945, it would seem, terror had become such a part of everyday life that it was regarded, at least by Josef Stalin, as a fit subject for a joke.

Was Stalin right in thinking that Roosevelt and Churchill would smile? He may have been, for two days later, February 13, 1945, 1400 British bombers and 2250 US bombers jointly launched what Canada's national newspaper (the 'Globe and Mail') described approvingly as part of the "terror bombing of ten German cities". The main objective the newspaper explained, was Dresden, selected on the grounds that it was a civilian target "jammed with refugees".

Of course the Allied forces did not initiate the use of terror bombing as a weapon. Its history goes back at least to General Franco's forces in the Spanish Civil War, a decade earlier. My point is that, little by little, terror had become an accepted weapon of war, with appalling consequences.

All this was to lead, later the same year on August 6th and 9th, to the incineration, without warning, of almost the entire civilian populations of two cities, Hiroshima and Nagasaki. This came as the culmination of three years of engineering effort by an international group of the world's leading scientists, under American leadership. Throughout those years the atomic bomb had surely been intended as an indiscriminate instrument for death and destruction.

Einstein – Peace Now! Reiner Braun and David Krieger (Eds.)
Copyright © 2005 WILEY-VCH Verlag GmbH & Co. KGaA, Weinheim
ISBN 3-527-40604-2

Ultimately, this led to an abiding revulsion against terror. The revulsion was clearly evident in the years following World War II.

It was evident in the founding of the United Nations to (as its Charter stated) finally "rid the world of the scourge of war". It was evident thereafter when hundreds of millions of the terrorized turned on the apparatus of terror, in one country after another. They dismantled the Soviet Union, tore down the borders of Europe, and embraced democracy. This is a process that is still continuing, and that, far more than terrorism, characterizes our age.

On this reckoning our revulsion over the attack on the World Trade Center is based not so much on the fact that it represents a harbinger of the future, as that it constitutes a fearful reminder of the past; a past we are determined we shall never be made to re-live.

The purpose of this reminiscence is two-fold. First to ask how this unprecedented evil of over half a century ago ever came to pass, and second where we must look in order to end it conclusively.

You may be questioning my remark that this was an "unprecedented evil". People have, after all, butchered one another throughout recorded history.

True. But they have never before had such ready means for slaughter. They could not treat one another, even had they wished, as insects to be killed by insecticide. That had to wait for modern science.

Nor, I believe, had human beings previously found such a powerful rationale for extermination. Science had provided that too, for the triumph of science seemed to represent the triumph of the machine.

Humans themselves had come to be regarded as machines, and machines are not subject to pity. Quite the contrary; they are subject to improvement by whatever means may work.

This, it should not need to be said, is a misreading of the message of science. Science is not a *deus ex machina*; a dangerous new form of religion. It is a very human and fallible pursuit. Nor is it amoral; it has morality at its core, since it teaches us devotion to truth. And since it is so evidently fallible, it teaches us humility (or should).

For all these modest avowals, science constitutes a major new strand in our culture. Most of the scientists who have ever lived, are alive today, because their profession continues to grow in importance.

The major visible impact of science is through technology, science's sturdy child. It is through technology that we are most conscious of science. But let us be clear that this child is not the parent. To have applied science, you must have science to apply.

Universities deal, as their name implies, with the universe of knowledge. They are rich in 'context', offering the ideal setting in which to explore connections. When those connections are surprising, they are called discoveries.

It is sensible to think of scientists as mapmakers. That innocuous simile has a lot to say about the culture of science, and about the functioning of a university. For it is impossible to make a map if one is not free to follow the contours of the landscape.

The freedom to follow a thought where it leads is not one we should take for granted. We in the universities would not have it if we were employees of government or industry. Government and industry have agendas. Their employees must further those agendas.

The freedom, instead, to follow a thought wherever it leads, is a freedom unique to the university setting. It is the university's strength, and it is an essential element in scientific discovery. We must guard it. And since freedom is to be valued, this is a reason that we value science. The respect for freedom lies at the heart of the morality of science.

It may surprise you that the university researcher's freedom is threatened. Surely everybody wants scientists to make discoveries? So, they do. But those who pay for the research, want to guide it. In doing so, they can easily impede it. In their desire to avoid waste, they then cause waste by insisting on programming the unprogrammable.

To bring these generalizations to earth let me say a word about my own scientific interests, which I share with a group of half a dozen students.

Our experiments have been designed to improve understanding of how molecules move in the course of chemical reactions. At first we studied luminescence from reacting gases at low pressures so that molecules newly born in chemical reaction, before they got beaten up in collisions, could signal to us their states of motion. We managed, with difficulty, to detect a characteristic feeble infrared glow. In short order, and much to our surprise, this became the basis for a large category of powerful infrared lasers.

But we shouldn't have been surprised. If you can make something feeble, then by adding the ingredient of knowledge you can quite likely make something powerful.

That research was performed some time ago. Today our work is no longer done in gases, but at the surface of crystalline silicon. The objective is unchanged – to find out how reacting atoms and molecules move – but the technique is different.

Using a sharp whisker of metal as a probe (a tool discovered elsewhere) we can see our reacting molecules individually. They constitute a sort of 'ink' of molecular-size dots, distributed over the silicon surface. What they do when gently tickled with light, is imprint themselves as patterns on that surface. At present we only partially understand the molecular dance that leads to these patterns. We hope to have the freedom to explore this systematically, in the coming years.

There is at the same time a competing interest, coming from the applied community. They would like us to use our molecular inks, and the methods for tickling them into print, in order to construct a printing press in which the print is of molecular size.

But, the goose of abstract knowledge must be nourished if it is to lay this egg. We don't even know yet if it is a golden egg. All we know is that it will not be, if we kill the goose.

Calls for freedom for the researcher to pursue basic science, are not calls for freedom from responsibility. University scientists must be held accountable, but to the right accountant; one who values new truth ahead of fulfillment of a plan, and new insights ahead of claims of cash value.

I am not at all one to discount wealth in a world where so many go hungry. But the greatest contribution science has made to the modern world is *not* through technology, important as that is, but through the example of the community of science. This is an extraordinary example of a highly competitive society that nonetheless puts truth ahead of personal advantage.

We are back to my earlier claim, that there is morality at the heart of science.

In considering what is true, science discounts nationality, race and ethnicity. It is sufficiently civilized that it does not merely tolerate dissent, but encourages it. It has no heredity ruling class, no formal government, no police and yet little crime.

It embodies much that is civilized. The fact that progress in science depends on humane values is often overlooked. The people who overlook it must be puzzled that scientists like Linus Pauling in the USA, or Sakharov in Russia, or Fong Lizhi in China have been leaders of democracy movements. Similarly they must be puzzled that science, like the arts, wilts and dies under dictatorships.

But I have already stressed how vital freedom of thought is to science. Scientists need and embrace freedom, since they must follow the trail of truth where it leads. The scientific community is tolerant since, in common with any democratic society, it feeds on dissent.

And yet the century past, with science in a position of dominance, has been one of dictatorship and savage intolerance. How can one reconcile these things?

The answer, as I have already hinted, is that the successes of science have been dangerously misrepresented. Science has been regarded, for example, as embodying an unassailable method of proof. There is no such method, and no such proof.

There is only the careful accumulation of evidence, leading sometimes, as in a court of law, to a conviction. A conviction that can at any time be overturned by the uncovering of new evidence. And most often is.

Scientific propositions, as I have already said, are not revealed truth. They are not edicts to be obeyed. They are stations on the path to truth.

There is no basis here, therefore, for the merciless jihads of Hitler and Stalin. Once debate was free, both of these movements languished. For it was terror, not truth, that supported them.

The same process of education will surely lead to the demise of the contemporary jihads, whose hold on human imagination is even weaker. Osama bin Laden, on the run and in hiding, is not Hitler or Stalin. He represents the last gasp of a messianic past in which the untutored were offered a magic route to happiness. This was such a shining path, they were told, that any means, however barbaric, could justifiably be used to clear it.

There is an error by which we could transform these relics of primitive thought, into a viable modern movement. This would be if we were to leave large segments of humankind without hope.

I have had something to say about the civilizing effects on science of a voyage of discovery. But those civilizing effects are only present

if there is a broadly shared vision in the community – as there surely is in the scientific community.

Louis XVI and the people of France, to give an awful example from
the past, had no such shared vision. The social collapse that resulted
first gave us the term 'Terror' as part of the political lexicon. That
should be an object lesson for all time.

Of course we can oppose terrorism by banning it, and should. We
can also ban war, which is terror writ large, and should. However, it
would be absurd to think that such bans can be effective in the absence of widespread popular support. There are not enough police in
the world. But popular support will be withheld so long as there is
rampant injustice.

Every year a million children die in some countries for lack of the
vaccines routinely administered in most countries. At the same time
ten million adults perish from easily preventable diseases, saving the
developed nations the one billion dollars it would take to rescue
them.

Terrorism is a crime and so, we are coming to realize, is neglect.
Recently, Nelson Mandela put it this way to a crowd of twenty thousand in London, England: "Massive poverty and obscene inequality
are such terrible scourges of our times (...) that they have to rank
alongside slavery (...)".

Had I made this comment about terrorism a year or two ago, I
might have been accused of blaming the victims for the terrorists'
crimes. But today we see the world more clearly. President Bush
spoke of the sources of terror in his January 20th, 2005 inaugural address, saying: "For as long as whole regions of the world simmer in
resentment and tyranny – prone to ideologies that feed hatred and
excuse murder – violence will gather".

In the same address he spoke of US "vital interests", in such a way
as to make it clear that these were no longer narrowly national interests, but as international as the pursuit of science has invariably
been.

His administration has yet to embrace the notion that international interests can only successfully be pursued by international
means. That realization must come. It will represent, it is true, a
humbling of the nation-state. But that is the consequence of enfran-
chising the people of all states. If, as we would wish, they are to have

a stake in their destiny, they must also have a stake in the world's destiny, free from the dominance of any single power.

The founders of the United States wrote a wonderful Constitution in 1786, balancing liberty against order. It promised "Justice, general Welfare and the Blessings of Liberty" to the people of that land.

What its framers actually had in mind, at that early date, were the white males of the land. Over the centuries, however, the promise was extended to an ever-widening circle. In our lifetimes, aided by the legacy of Einstein and the civilizing influences of science, this splendid promise is about to be extended to its proper beneficiaries – humankind.

John Polanyi

John C. Polanyi was born in 1929. He is a faculty member in the Department of Chemistry at the University of Toronto, Canada. His research is on the molecular motions in chemical reactions in gases and surfaces. He won in 1986 the Nobel Prize for Chemistry for his contribution concerning the dynamics of chemical elementary processes.

Einstein's Heirs: Szilard and Sakharov

Dudley Herschbach

In his book, *Science and Human Values*, Jacob Bronowski describes his stunning experience in visiting Nagasaki in November, 1945, to assess the damage wrought by the atomic bomb. He urged an action:

> When I returned from the physical shock of Nagasaki,...I tried to persuade my colleagues in governments and in the United Nations that Nagasaki should be preserved exactly as it was then. I wanted all future conferences on disarmament, and on other issues which weigh the fates of nations, to be held in that ashy, clinical sea of rubble. I still think as I did then, that only in this forbidding context could statesmen make realistic judgments of the problems which they handle on our behalf. Alas, my official colleagues thought nothing of my scheme; on the contrary, they pointed out to me that delegates would be uncomfortable in Nagasaki.

Sixty years later, mankind remains addicted to war. Modern prophets, many scientists among them, have cried out against the vast resources squandered on weapons and the appalling massacres committed in the name of patriotic, religious, or tribal loyalties. But like the ancients, prophets for peace today are still typically ignored or reviled as impractical visionaries.

Yet scientists have a special responsibility to join in efforts to overcome complacency and despair, "to light a candle rather than curse the darkness." We are privileged to take part in probing the wonders of nature, and thereby likely to be keenly aware of inspiring prospects as well as ominous dangers. In this essay, I want to pay homage to Leo Szilard and to Andrei Sakharov, scientists with humanistic passions for peace akin to that of Albert Einstein. Both had prominent roles in the development of nuclear weapons. Both tried mightily but in vain to persuade politicians and the public to avert a nuclear arms race. Both left political testaments of enduring value. These support what I deem to be major lessons that should be drawn from history since Nagasaki.

Einstein – Peace Now! Reiner Braun and David Krieger (Eds.)
Copyright © 2005 WILEY-VCH Verlag GmbH & Co. KGaA, Weinheim
ISBN 3-527-40604-2

Szilard: The Voice of the Dolphins

An émigré to the United States from Hungary, by way of Germany and England, Leo Szilard (1898–1964) was among the first to conceive of a nuclear chain reaction and to recognize what it would mean for the world. As a student in Berlin, where he received his Ph.D. in 1922, Szilard had adopted Einstein as a mentor. Later they collaborated on several inventions, and even patented a novel refrigerator. At the University of Chicago, Szilard developed with Enrico Fermi the first self-sustaining nuclear reactor. Soon after, Szilard proposed and helped draft the historic letter sent in 1939 by Einstein to President Roosevelt, which urged undertaking the atomic bomb project. In 1945, Szilard led the small group of physicists who opposed dropping atomic bombs on Japanese cities. He helped launch the Pugwash conferences in 1957; by bringing together Western and Soviet scientists, these had a significant role in promoting treaties to restrain nuclear weapons. Szilard also founded, in 1962, a political action committee, then a novel idea. This is the Council for a Livable World, which has helped elect more than eighty US Senators committed to work against the arms race.

In his efforts to foster sensible initiatives, Szilard wrote a series of fanciful stories. Most striking is one in which he recast the ancient tale of the Delphic Oracle into a modern allegory, titled *The Voice of the Dolphins*. In this fable, published in 1961, Szilard speaks from the perspective of a historian 25 years later. He describes how in 1963 a joint Russian–American Biological Research Institute came to be set up, "having no relevance to the national defense or to any politically controversial issues." It was located in Vienna and staffed by some of the most able young molecular biologists from both countries. Molecular biology was then a most exciting new research frontier. Thus it was startling when the first scientific papers to come from the Vienna Institute turned out not to deal with molecular biology but with the intellectual capacity of dolphins.

Soon it was established that dolphins were indeed far smarter than humans. Within a few years, the dolphins had mastered known science and shown great creativity in devising key experiments that led to dramatic advances in molecular biology. Five successive Nobel Prizes for medicine were awarded to the Institute for those advances, greatly enhancing the prestige of the dolphins. The Institute also produced a bioengineered form of a common algae, which it patented and marketed under the name Amruss. This was marvelous stuff, a source of cheap food with pleasant taste and excellent nutritive qualities. Moreover, Amruss also proved to markedly depress human fertility.

Amruss royalties made the Institute rich. It expanded by recruiting social and political scientists to work with the dolphins. It also purchased television stations in cities all over the world. These stations developed a major program for discussion of political problems, called "The Voice of the Dolphins." The program did not advocate particular solutions but was devoted to clarifying the real issues and indicating options and novel approaches suggested by the dolphins. Aided by their prestige and by judicious use of Amruss investments, these approaches proved successful. By 1988 a general disarmament is achieved. However, the Institute is disbanded after a virus epidemic kills all the dolphins. The staff departs to new research institutes, set up in the Crimea and in California, but does not equip them with dolphins. In his final paragraph, Szilard concludes:

> There were, of course, those who questioned whether the Vienna Institute had in fact been able to communicate with dolphins and whether the dolphins were in any way responsible for the conspicuous achievements of the Institute...It is difficult to see, however, how the Institute could have accomplished as much if it hadn't been able to draw on considerably more than the knowledge and wisdom of the Russian and American scientists who composed its staff.

This ironic twist is the hopeful moral. Szilard's message is that humankind can marshal its own wisdom to overcome tragic folly. But we must develop a mutual trust in our capacity and conviction that it can and must be done. Szilard would surely have been astounded by the abrupt political transformation of Eastern Europe in 1989. It occurred about when he imagined, although in a way unanticipated by him or any scholar or political leader. Yet it exemplifies an essential aspect of Szilard's fable, with added irony. TV proved a key factor, by showing the Western world to people behind the Iron Curtain. Without benefit of any formal program, TV generated a collective awareness that served as the Voice of the Dolphins.

Sakharov: Courageous Voice for Reform

Andrei Sakharov (1921–1989) was born into a family of Russian intellectuals. His father was a physics teacher and author of popular science books and teaching texts. During World War II, Sakharov worked in a munitions factory. At its end, he began graduate studies at the Physics Institute of the Academy of Sciences in Moscow. He received his Ph.D. in 1947 for fundamental theoretical work in nuclear physics. Recruited into the Soviet hydrogen bomb program, and imbued with what he termed "a war psychology," he worked assiduously on the project, came up with key ideas, and became a chief designer of the Soviet H-bomb. He soon came to realize the immensity of the threat posed by such weapons and in 1958 urged that testing of nuclear bombs be halted – to no avail. As the Cold War and arms race accelerated, Sakharov "felt a growing compulsion to speak out" because I shared the hopes of Einstein, Bohr, Russell, Szilard, and other Western intellectuals that notions [of open society, convergence, and world government], which had gained currency after World War II, might ease the tragic crisis of our age.

In June, 1968 Sakharov published a bold manifesto, *Progress, Coexistence, and Intellectual Freedom*. He urged disarmament and an end to the Cold War and proposed steps to remaking the Soviet Union. Among these were instituting democracy, including freedom of expression and abolition of censorship, scientific exchanges, reform of economic and social systems, curbs on the power of security forces, limiting the defense against external threats, and full disclosure of crimes of the Stalin era. He also argued for an evolutionary coexistence of socialism and capitalism. His manifesto coincided with the "Prague Spring," and Sakharov endorsed its theme of "Socialism with a human face." In July, the New York Times brought Sakharov to world attention by devoting three pages to printing his *Progress*. Western reaction was strongly positive and widespread; within a few months, the English edition of *Progress* had sold 18 million copies. The Soviet response was expressed in swift denunciation of Sakharov and the invasion of Czechoslovakia in August to suppress the "Prague Spring."

Sakharov was undaunted. He continued to campaign against nuclear testing and ideological distortions of science as well as to urge drastic economic and political reforms and to protest the persecution of other dissidents. For his *Progress*, Sakarov had chosen an epigraph from Goethe's Faust: "He alone is worthy of life and freedom who each day does battle for them anew." Years later, in his *Memoirs*, he commented on this choice: "The heroic romanticism of these lines echoes my own sense of life as both wonderful and tragic."

Even after Sakharov received the Nobel Peace Prize in 1975, Soviet officials remained implacable and did all they dared to slander and persecute Sakharov. When he opposed the Soviet military invasion of Afghanistan, he was banished in January 1980 to Gorky. For six years there, he was vigorously harassed by the KGB. Then, in December 1986, came the dramatic phone call from Mikhail Gorbachev, inviting Sakharov and his wife, Elena Bonner, to return to Moscow. There he continued voicing his appeals. After the abrupt political upheavals that ended the Soviet Union and reshaped Eastern Europe, a few months before his death, he was elected to the new Soviet Parliament. Many of the reforms he had long urged became official objectives of *perestroika*.

At a memorial commemoration of Sakharov held in Cambridge, I was much impressed with an aspect emphasized by Elena Bonner. She pointed out that, prior to his election to the new Soviet Parliament, ordinary citizens had a viciously distorted notion of Sakharov and his ideas. For 30 years, they had known only the caricature created by the official press. But the sessions of the Soviet Parliament were shown in full on TV. To the amazement of Elena Bonner, within two weeks the public view of Sakharov had changed to one of admiration and trust. This again exemplified the power of TV to serve as the Voice of the Dolphins.

Beyond Nagasaki

Only about a quarter century after Nagasaki, in the Vietnam war, the world's richest and militarily most powerful nation was defeated, after a long, bloody struggle, by a country with a population tenfold smaller and a per capita income at least seventy-fold lower. In a sequel a decade later in Afghanistan, the world's other military superpower was likewise defeated. In both cases, lack of understanding of the indigenous culture proved to be a severe handicap for the invaders. As in earlier history, these defeats of proud and powerful nations had profound impact, still unfolding, on their internal poli-tics. In contrast, the astoundingly rapid collapse of the Iron Curtain and the Berlin wall and the subsequent political restructuring of Eastern Europe were nearly bloodless. That huge transformation was not imposed from without, but resulted from internal political pres-sures that built impetus for spontaneous reform.

Sixty years after the bombing of Nagasaki, the likelihood that a nu-clear weapon will again be exploded on a city seems higher than ever. This ghastly prospect is the consequence of failure to curb prolifera-tion and terrorism, now spurred by the Iraq war. The situation is very different from that faced twenty-five years ago; then, in the name of Cold War deterrence, more than 50,000 nuclear bombs were de-ployed, enough to destroy every sizable city in the United States and the Soviet Union several times over. Yet much that is still relevant can be found in a collection of essays, *The Final Epidemic*, published in 1981 as a joint project of the Council for a Livable World Education Fund and Physicians for Social Responsibility. Most valuable today are current issues of the *Bulletin of the Atomic Scientists*. Launched soon after Nagasaki, with Albert Einstein and Leo Szilard among the original sponsors, it is the premier forum for global security analy-sis.

As the world will need more and more energy generated from nu-clear reactors, opportunities for terrorists to acquire fissionable ma-terial will grow. An excellent recent book, *Megawatts and Megatons*, treats both power reactors and weapons, including extensive discus-sion of the crucial need to enhance security measures and arms con-trol. In this era, a much-heightened level of international cooperation has become essential. At present, this is impeded by refusal of the United States government to ratify either the Comprehensive Test Ban Treaty or the Nonproliferation Treaty. Ratification would help create an attitude of shared concern. That attitude is important for obtaining reliable intelligence, the most vital safeguard against ter-rorism.

Szilard's visionary fable suggests a pragmatic way to transcend such myopia of individual governments. A worldwide TV/radio/web network could be established to function like his Voice of the Dol-phins. It would operate under the auspices of the United Nations, or perhaps a coalition of nongovernmental organizations, broadcasting in many languages. It could be funded by a tiny tax on defense budgets. It would make use of existing broadcasting capabilities where they exist and create them elsewhere. It would also develop, subsidize, and distribute where necessary receiving equipment to people lacking them. In addition to regular programs dealing with issues of global concern, including medical and environmental as well as security issues, the network could broadcast news, cultural and educational programs, extending from elementary to university level. Much suitable programming could be adapted from existing facilities but would have to meet criteria of objectivity and broad perspective, like those specified by Szilard. The role of his dolphins would typically be performed by expert panels convened by science academies, heirs of Einstein.

An apt benediction was given by Andrei Sakharov in accepting his Nobel Peace Prize:

> (...) We should not minimize our sacred endeavors in this world, where, like faint glimmers in the dark, we have emerged for a moment from nothingness (...). We must make good the demands of reason and create a life worthy of ourselves and of the goals we only dimly perceive.

References

Jacob Bronowski, *Science and Human Values*, Rev. edn., Harper & Row, New York, 1965.

William Lanouette with Bela Silard, *Genius in the Shadows: A Biography of Leo Szilard*, Scribners, New York, 1994.

Andrei Sakharov, *Memoirs*, Knopf, New York, 1990.

Ruth Adams and Susan Cullen, (Eds.), *The Final Epidemic: Physicians and Scientists on Nuclear War*, Educational Fd. For Nuclear Science, Chicago, 1981.

Richard L. Garwin and Georges Charpak, *Megawatts and Megatons: A Turning Point in the Nuclear Age?*, Knopf, New York, 2001.

Dudley Herschbach

Dudley Herschbach, born in 1932, is a Mathematician and Chemist. His research is devoted to methods of orienting molecules for studies or collision stereodynamics, means of slowing and trapping molecules in order to examine chemistry at long deBroglie wavelengths, reactions in catalytic supersonic expansions, and a dimensional scaling approach to strongly correlated many-particle interactions, in electronic structure and Bose–Einstein condensates. For his work he received the Nobel Prize for Chemistry in 1986.

Making it Happen

Jakob von Uexküll

1. Global Values and Global Stability

At the start of the new millennium the world needs an independent body to inject a missing ethical dimension into the conduct of national and global affairs. We need a clear and sustained voice that expresses our values as world citizens, rather than just as global consumers. Our key challenge is not a 'values vacuum', but that widely agreed values are not being acted on.

The *World Future Council* (WFC) is being set up to ensure that ethical, long-term thinking becomes central to the debate about our common future.

The challenges that humanity now faces are historically unique, both in their globality and their vast time horizon. For the first time in our history we are not just affecting the next 100 years by what we decide – or don't decide – today, but millennial or even geological time spans. We have the power to severely affect present and future generations, yet we seem incapable of dealing with the global and long-term impacts of our actions. The integrity of our planet is being damaged by the impacts of a global consumer society, yet despite the growing clamor for change, alternatives are rarely explored, leaving many of us feeling increasingly angry and frustrated.

We simultaneously celebrate and lament our 'mastery' over nature. We seem to be part of some automatic, seemingly unsteerable process called progress, but our modern experiment of putting scientific and economic freedom first, and then bringing in ethics to deal with the consequences, is not working.

Our greatest problem is not poverty, environmental collapse or terrorism – it is our failure to respond to the great challenges of our time, despite having the knowledge and power to do so.

Einstein – Peace Now! Reiner Braun and David Krieger (Eds.)
Copyright © 2005 WILEY-VCH Verlag GmbH & Co. KGaA, Weinheim
ISBN 3-527-40604-2

The World Future Council aims to rise to this challenge. It seeks to be a powerful global voice that appeals and responds to our basic human values.

The power of the Council will be primarily ethical – but its significance should not be underestimated. As a voice of Global Stewardship, it will be able to provide valuable guidance, and become a powerful agent for change. By reflecting our common values and responsibilities for the present and the future, it will give significant impulses both to research and to action.

The Council will provide an ethical audit on important decisions. It will be a 'community of conscience', with the authority, conferred by the stature of its members, to pronounce on the great issues of our time. It will be legitimized by the quality of its work and its global membership. As a permanent forum, it will act as a global conscience by speaking up for our values, rights and responsibilities as citizens of the world.

It will listen, study and speak out at regular public sessions, nationally and internationally. It will aim to strengthen other initiatives that are seeking to foster global interdependence and responsibility. It will oblige decision-makers to look beyond the political and economic advantages of any given project, and to ask: 'is it good for people and planet, now or in the future?'

The Council will analyse and quantify the yawning action gaps on important issues such as climate change, deforestation and global inequality, and push for reforms to bridge the gaps between what is currently being done and what actually needs to be done. It will work closely with policy makers, as individuals and through parliamentary groups such as the E-parliament, a new initiative that will electronically link the world's democratically elected MPs. It is envisaged that Council members and MPs will hold joint hearings to formulate and adopt concrete proposals to campaign for in their national parliaments and other relevant bodies.

2. The Council

The World Future Council will have a core membership of up to 100 persons, consisting of respected individuals from across our planet – wise elders, 'heroes', 'whistle-blowers', 'best practice' pioneers, and youth leaders. They will meet annually, addressing key issues of our time. A smaller executive committee will meet more frequently.

The challenge today is not to create 'better' humans but to choose which values our societies should prioritize, respect and protect, for instance;

- Sanctity of life, or commercialization of its building blocks?
- Citizen, or consumer values?
- The well being of the poor, or the primacy of economics?
- Lifestyles of modest sufficiency, or rampant consumerism?
- Local rights, or global monopolies?
- Reciprocity and solidarity, or competition and profit?
- Human happiness and spiritual well being, or material growth?
- The interests of future generations, or only our own?
- Mastering, or giving in to our desires?
- Cultural diversity, or global media dominance?

The World Future Council Initiative is now working to develop the WFC, and the practical aspects of this are summarized below. The Initiative has already been endorsed by many leading thinkers and actors across the world, and their names are listed on our website www.worldfuturecouncil.org. We will be pleased to add your name and to have a few sentences of support from you.

3. The Underlying Issues

We have unprecedented power over and responsibilities for both present and future generations. But, blinded by our successes, we have lost our place in the larger story of life as we listen to 'experts' who tell us that society is only a jumble of conflicting interests filled with human 'robot vehicles blindly programmed to preserve the selfish molecules known as genes' (Richard Dawkins).

It is claimed that we live in a 'values vacuum', or that people across the world have irreconcilable differences in their value systems. There is little evidence for such claims. Researchers at the Institute for Global Ethics and others have found a remarkable agreement on values and value priorities across continents, faiths, cultures, social classes and religions. This basic consensus overrides diverse world views and is shared by believers and non-believers of very different backgrounds and countries.

A global citizens' community with common values does not have to be laboriously created. It already exists. But we lack institutions to back up these common values. The key problem is not 'the way we think', but which of our thoughts are respected and acted on by the institutions of power in our societies. We do not face a values gap. We face an action gap.

At a time when the world craves visionary leadership, many of our political leaders have become prisoners of an autistic economic fundamentalism that discounts the future and undermines the values and institutions on which the urgently needed transformation of our societies can be built. They no longer represent us as citizens, but only in our narrow capacity as consumers. They have focused their energies on giving the international legal protection of corporate profits a higher standing than basic human, social and environmental rights. They therefore face cynicism, disinterest and, sometimes, violent opposition by all those who feel alienated from a system that offers such dismal choices.

If we want to avoid rising levels of global conflict and environmental stress, we need to re-cast the debate on our future in moral terms, and impose our citizen values on our economics, instead of vice versa. Economic doctrine and its practical policies, i.e. the global capitalist 'growth' model, operate at a fundamental level in contemporary human activities. Its theories have usurped our value systems, cultures and traditions to such an extent that most human, social and environmental problems are now seen as economic. The dominant question is: 'Can we afford it in the short term?'

The orgy of consumerism over the past forty years has not made the global rich any happier. More and more people feel increasingly uncomfortable, fearful and frustrated with a system that sacrifices deeply held values for a single goal – global consumerism – which is in direct conflict with maintaining a livable planet.

Our leaders deplete our ethical, social and natural capital, 'sell the family silver', monetarize our non-market wealth – and ask to be applauded for this unprecedented 'wealth creation'. They sacrifice the children of the poor on the altar of their accounting practices and banking regulations. They embrace the commodification of life through genetic manipulation, applying a false reductionist and mechanistic mindset to living systems. This economic system is fast becoming humanly, socially and environmentally unaffordable.

In many areas it may already be too late for an orderly transition. Global oil reserves have been exaggerated for political and commercial reasons. Production is expected to peak in a matter of years, severely destabilizing a world order built on cheap oil. The United Nations Environment Programme, UNEP, calculates that climate change could bankrupt the global economy in less than a lifetime unless drastic measures are taken.

We need to detrivialize the global debate and speak some simple truths. Much of what is currently politically acceptable is in fact international and inter-generational commercial and financial terrorism. His Holiness The Dalai Lama writes in his latest book: 'I find it difficult not to suspect that, by means of international debt and the exploitation of natural resources at relatively low cost, the wealth of the rich is maintained through neglect of the poor.'

Allowing ever-increasing CO_2 emissions causing global climate change is a crime against humanity – as are policies forcing the poor to pay dubious debts to the rich while ignoring the historical and ecological debts of the North. Those responsible must be held accountable.

Our ancestors ensured that private corporations, which acted against the common good, could lose their privileges. In the 18th century the British government could dismantle any commercial enterprise 'tending to the common grievance, prejudice and inconvenience of His Majesty's subjects'. Until quite recently similar legislation existed in the USA.

The consensus, which permitted such disincorporation, was destroyed by a coup d'état of a rich minority – and their servants in politics, the judiciary and the media – who now want to globalize their rules and rule. They claim that we are all united only by common greed. But this is not so. When we are addressed as consumers we respond as such. When nothing else is offered or demanded, shopping becomes the principal cultural expression of 'modern' societies.

Our common values as citizens of the Earth are currently drowned out by the cacophony of commercial speech. In our societies the loudest voice is that of advertising. Starting with pre-school children, it aims to foster a culture of permanent dissatisfaction, immaturity and irresponsibility. The youth delegates at the first State of the World Forum (1995) described the consequences:

> "We currently face a global crisis of the spirit in the search for meaning. As our confidence and self-esteem decline, the value of friendship, family, society, trust and respect begins to lose the battle against selfishness and the pursuit of material gain. It is difficult to know what to believe in these days"

The most serious threat today is not the continuation of present policies – which will soon be impossible. It is the collapse of our societies as our leaders lose their credibility and are replaced by preachers of intolerance and obscurantism in a reaction against market fundamentalism.

New ethical leadership is now needed to counter this global emergency. Twelve years after the Earth Summit in Rio, and two years after the World Summit on Sustainable Development in Johannesburg, the overall signs still point in the wrong direction. There has been a major effort to water-down 'mutual accountability' for dealing with global ills. The debate on shared responsibility is receding and the poor global majority is losing faith in democracy. Even limited agreed reforms, such as the UN Millennium Goals, are unlikely to be implemented in the proposed time horizon.

These are not my conclusions but those of two of the highest-placed insiders in this process, the heads of the World Bank and the United Nations Development Programme, UNDP. We can either fail, and go down in history as accomplices in monstrous crimes, or we can act now.

4. A Rising Tide of Anger

We are outraged by 'Christian' leaders who insist that the world's poor pay compound interest to the rich even when it costs the lives of their children – in defiance of the command of every religion! We fear for the health of our children, knowing that for many years now mother's milk in many countries is too contaminated to be marketable. We fear the judgment of our grandchildren when the richest nation on earth declares it cannot afford the costs of preventing global climate destabilization, estimated at one per cent of Gross National Product.

We feel hurt to the core of our being by the daily horrors of animal experimentation and of agribusiness: of the 250 million cattle, pigs, horses and sheep transported across Europe every year, many for up to 20 hours without rest or water, 25 million are dead on arrival. We are deeply affected when nature is dying and even the experts admit being 'scared' because no-one knows why – as when large scale fish deaths occurred in the Baltic recently.

We feel diminished by the mechanical dogma foisted upon us, which portrays nature as a mere machine to be manipulated at will. We feel outraged by corporate control over the genetic blueprints of life, and worry about future generations thinking of life as a human invention, with no boundaries between the sacred and the profane. (In the USA, couples can now order a designer child over the internet – eggs, sperm as well as surrogate mothers.)

We feel cheated: by the dismissal of deeply held values in the name of economic efficiency; by the promises of a 'micro-millennium' and a 'silicon civilization' of freedom and leisure, against the reality of increasing working hours, and stress or long-term unemployment and exclusion; by the promise of a 'global village' against the reality of a return of the brutal competitive world of the 19th century, in which deskilling and adversarial money bargaining break down trust and community; by the 'ecological aggression of the North against the South' (Klaus Toepfer); by the continued nuclear 'weapons' race, with its potential to poison the earth and its inhabitants for generations to come; by a science that produces toxic cocktails poisoning our water, soil, air and bodies.

I say 'we' because in my experience anger and outrage is a common reaction from those offered the opportunity to respond. It is reflected in reports from all over the world, whether it is 85 per cent in a UK poll demanding the protection of local production from enforced globalization, or the 'citizen juries' in Latin America, overwhelmingly rejecting GM foods – after hearing both sides of the argument – and other aspects of corporate-driven globalization.

Many react to the ecological–cultural crisis with despair, searching for scapegoats, becoming addicted or falling ill. The rising tide of intolerance as well as the epidemics of drugs, depression and mental illness in the industrialized countries indicates that we are on a path that is destructive not just to our planet but to our societies and to ourselves.

5. Economic Totalitarianism

Over 20 years ago, German chancellor Helmut Schmidt found it necessary to justify his country's foreign aid as creating future customers for German exports – a far cry from the poet Hjalmar Gullberg's belief that 'a hungry human being less is a brother/ sister more.' Which reasoning resonates more deeply with us?

Today the whole world is a third-world country with a poor majority, and 'the gated communities' of the rich are less and less effective in keeping them out. Either the poor are able to live decently at home or they will come to the rich world in ever-increasing numbers. Sending them back will not be a realistic option. In global market societies economic migration is rational, especially when the economic policies imposed on the poor are failing.

Individualistic market ideology teaches that all failure is personal. So instead of staying and fighting injustice at home, many poor people take the rational decision to try elsewhere. Today 85 per cent of young North Africans want to emigrate to the European Union: What will the EU do when the numbers coming increase 10-fold or 100-fold? Trying to keep them out will only increase the number of those who join the most visible global competitor of market fundamentalism: fundamentalist Islam.

All over the world societies that have to educate their children in opposition to the dominant cultural message, i.e. consumerism, are becoming unsustainable: in Thailand, for instance, the Buddhist 'principles of virtuous existence' are no longer being taught beyond primary school. In the USA the sales of anti-depressants to children from the age of two are growing rapidly. In the UK 50 per cent of children under 16 report being 'stressed out', disorientated by the rapid, disruptive changes in a market-driven society.

Today the opposition to the rule of 'economic man' – with its incomplete understanding of human nature – is growing rapidly and it is increasingly values-based. The objections to feeding herbivores like cows with ground-up dead animals were mainly ethical, as the health risks (BSE) where not yet known. But these objections were rejected as 'unscientific', as are the objections to GM foods today. As a result, 'big science' is increasingly seen with suspicion, as an extension of big business, and no longer to be trusted.

Scientists whose findings disagree with the application of mechanistic thinking to living systems are ignored or dismissed. This was the experience of my biologist grandfather 80 years ago, and of the biochemist Michael Behe – author of 'Darwin's Black Box' – in recent years, to mention just two. Consumerism is the 'rational' reaction to this depressing story: if the material is the ultimate reality, then maximizing our material possessions seems a logical way to drown our existential despair. The shopping mall becomes the symbol of a community's modernization.

'Governments should not hinder the logic of the market', says Tony Blair – forgetting that the cornerstones of this market (e.g. limited liability, fractional banking, financial globalization and the WTO), are state-created and dependent on it for their existence. Markets make good servants but bad masters.

By overriding values-based objections in the name of science and economics, we risk 'barbarization from within' (Lewis Mumford), for such objections go to the heart of our understanding of what it is to be human. The present global order and economics are based on a materialistic worldview that implies that we are ultimately just products of 'accidental collocations of atoms' (Bertrand Russell). This is the modern story we are born into, the 'truth' of a meaningless universe.

6. Shared Values

We need a new global and inter-generational social contract that reflects our need for balance and sufficiency.

International commissions have shown a remarkable consensus when their members from very different backgrounds and belief systems are asked to focus on the needs of the planet and future generations. It is often claimed that European or Western values prioritize individual rights while Eastern values put communal rights first. But the realities on the ground are different. Over 20 years ago I set up the Right Livelihood Awards, presented annually in the Swedish parliament, with many recipients from other continents. The only times I have ever been accused of not understanding Asian or African values were when the Indonesian and Nigerian dictators complained about awards to human rights activists in their countries.

We all have many different values – we are potentially both devils and angels – but in normal circumstances our value priorities are quite similar. Our ancestors relegated waste and conspicuous consumption to special occasions and feasts – otherwise our planet would probably be uninhabitable by now. Extreme selfishness is usually seen as acceptable only in exceptionally threatening circumstances. People are honored for what they do for others, for the community. Individualistic greed is frowned upon – except in our modern Western culture, where such behavior is celebrated as normal and commendable, for it is what the market demands. And as the chairman of the Federal Reserve Board Alan Greenspan said recently: 'Markets are the expression of the deepest truth of human nature and will therefore ultimately be correct'. But is consumerism really our 'deepest truth'?

The founder of Transpersonal Psychology, Abraham Maslow, pointed out that it is difficult to practice higher values like love, generosity and solidarity in a society whose rules, institutions and information streams are set up to promote lesser human qualities. Most of us are not heroes and will follow the values of power even if we feel uncomfortable doing so. If we are really united only by greed and otherwise divided by deep value differences, then our situation is hopeless. For where is the 'new ethics' and 'new way of thinking', which is increasingly called for, going to come from? What is it to be based on?

But if key human values are shared globally, while the ways we interpret, honor and live them reflect the power relationships in our society, then there is a way forward, because every one of us can contribute to changing existing norms simply by challenging them.

It is sometimes enough just to ask a different question. For instance, we are told that Americans think that the USA pays too much in foreign aid. The 'Americans Talk Issues Foundation' polled Americans on how much aid they thought their country actually pays as a percentage of GDP. Average answer: 18 per cent. What would be fair? Average answer: 5 per cent, i.e. about 50 times more than the US currently pays!

Cutting aid to the poor – and then complaining about economically motivated immigration – is part of the same compartmentalized thinking that worries about increased Chinese consumption, but also about the Japanese not consuming enough. The same mindset wants to lower the retirement age to reduce youth unemployment, but also to raise it to reduce costs. The same mindset worries about the huge level of debt, but worries even more about the consequences if it falls.

A democratic system that keeps us busy choosing between different electricity providers, but does not allow us a choice on issues affecting deeply held values, will not survive. As the Pintasilgo Commission noted a few years ago, 'carrying capacity is a function of caring capacity'. A political system that fosters greedy individualism is destroying 'caring capacity' at a rate that will soon not just threaten immigrants but the very cohesion of our societies.

7. Service and Balance

The primary goal of any sustainable human endeavor, including business, must be service and balance, not profit and growth, as our ancestors knew. We need a serious debate about who we are and who we want to be. What sort of relationship with our natural (outer) environment is attuned to our human identity, i.e. our inner environment?

We are the first generation to affect even the Earth's climate and the last that does not have to pay the price for this. The market increasingly crowds out our core survival activities as economically worthless and too time-consuming. It is now well established that the key factors that affect happiness most are mental health, satisfying and secure work, a secure and loving private life, a safe community, freedom and moral values. The political implications of this are devastating, for the current global 'reform agenda' points in the opposite direction – of more insecurity ('mobility', 'flexibility') in return for ever more things. (Its inability to cope with the Japanese, who have decided that they have enough things, illustrates its perversity.)

Increasingly this agenda is facing a 'revolt from the center'. International youth conferences all over the world have been demanding 'commerce-free space'. In local referenda large majorities, in countries as diverse as Germany and Brazil, reject key aspects of global economic integration. This refusal makes sense in a world of massive unemployment, in which the most secure jobs are actually created by small and medium-sized enterprises that produce for regional markets. They create jobs at a fraction of the cost compared with those few created by footloose, global companies that roam the earth like the insatiable 'hungry ghosts' of Buddhism.

It is fashionable today to complain about 'too much government' – while promoting economic globalization that will necessarily mean even more government to protect us from its side-effects. The alternative of more 'regulated self-regulation' must include the right of every community to regulate the extent of its global market integration, e.g. through local content and technology-transfer requirements. Such policies have helped countries like the USA, Scandinavia and the Asian 'Tigers' to prosper in the past – but are now outlawed by the World Trade Organisation.

It is increasingly obvious that the ruling ideological paradigm with its mixture of modernist consensus and post-modernist relativism (money rules, science has the truth, everything else is a matter of opinion) is not serving us well. To escape its limitations we need to create the right institutions to guide us on our unprecedented journey, by helping us expand our sense of identity and recognize that we have the knowledge, technology and wealth to create a fair and future-compatible (sustainable) global order in tune with our highest values.

8. Global Responsibilities and Moral Imperatives

A forum with the ethical and intellectual authority to guide this deep cultural transformation is urgently needed. No other institution is presently filling this gap. The most glaring failure is probably that the majority of religious leaders today lack the courage to speak out against global consumerism and the destruction of biodiversity, and seem to have lost their ability to engage in moral discourse.

The task facing us is huge, but we do not start from zero! The building blocks of a workable global future already exist. Putting them together requires a convincing vision that captivates imagination and inspires action – both of those now excluded, and worried about their family's next meal, and of those who can penetrate and help redirect the global power system towards a just and sustainable order.

The alternatives are becoming clear. Either we are able to respond with the required psychological and spiritual maturity to the unique challenges facing us; or the revolt against market domination of our lives will spread and turn even nastier. History has several examples of mighty materialistic societies collapsing, to be followed by centuries of intolerance, deskilling, economic collapse, wars and terrorism. But, for the first time, if such a collapse occurs today, it will affect the whole world.

9. The World Future Council

Speaking for shared human aspirations, values and responsibilities, the Council can become a powerful agent for change. It will help us reclaim our minds and sensitivities, eroded by consumerist brainwashing. It will build on the invaluable work that has already been done by the international community to define our rights and duties as planetary citizens.

The Council will not attempt to 'represent' others, but rather to express and manifest common values and goals – as citizens taking responsibility for the future. It will aim to be a catalytic force that crystallizes the moral/intellectual offensive against the ideology of 'moneytheism'. It will restore confidence in our power to change by inspiring and advancing our vision of possible alternative futures. It will encourage those who feel voiceless, alienated and excluded, and provide leadership and protection for moral courage.

The WFC will help prepare us for the tough decisions ahead by fostering an ethical culture and nurturing mindfulness about the consequences of our actions. It will highlight the ecological, human, social and institutional costs of current 'growth'.

Speaking the language of values, the Council will provide an ongoing reminder of the daily betrayal of future generations. Its standing and moral power will grow as the gap it fills in the architecture of global governance becomes apparent. It will stimulate the creation of national, regional and local Future Councils, and serve as a resource and reference-point for their work. It will ensure that the invaluable work done by international commissions during the past decades is connected, built on and given a permanent voice, instead of being filed away and ignored.

While WFC members will serve in a personal capacity, the aim is to include recognized leaders and representatives of different geographical areas and sectors of society who have shown an awareness and understanding of global values. The Council will be complemented by thematic advisory commissions from civil society, politics, academia, culture, business, etc., dealing with key global issues.

The World Future Council will hold the long-term vision, and ensure that our elected policy makers have access to the best practical visionaries and long-term thinkers when tackling global problems. (The Australian Green Senator, Bob Brown, suggests that above the entrance to every parliament should be the words: 'Will people 100 years from now thank us for what we are doing here?'). The Council could evolve into an elected Earth Senate. Several traditional societies had a 'Council of Seers Into The Future' whose voice was respected when day-to-day decisions were taken.

10. Creating the WFC

This project is currently co-ordinated by a small international planning committee (IPC). The founding organizations are currently The Right Livelihood Awards Foundation, EarthAction, BAUM – the Association for Environmental Management, The Global Challenges Network, and the UK Schumacher Society.

The WFC Initiative has held preparatory meetings with key thinkers and activists from all continents in Salzburg (Austria) and Tenerife (Spain) – jointly with the e-Parliament Initiative – at the invitation of the regional authorities.

We are aware that at present the WFC Initiative team and office locations are not globally representative. We are working to correct this as a matter of urgency, as our resources permit. This project has been endorsed by many personalities and organizations from the South who wish to help build a fairer global order.

Current WFC Initiative activities include:

- Approaching a number of cities and regions, asking them to host all or part of the WFC secretariat, if they can support WFC operations during the 3–5 year initial phase with the help of local sponsors. (Afterwards, the WFC will have become a key part of the global governance structure, able to attract major institutional funding.)
- Writing to 8500 Civil Society Organizations in 170 countries, asking for their support and feedback. They have also been asked to propose names of respected individuals as possible WFC members from their country.
- Contacting 15,000 (out of approx. 25,000) democratically elected national Members of Parliament with the same requests. They will also be asked to indicate on which issues they want to work with the WFC.
- Producing several publications, including this book, and the first 'World Future Report', which will summarize the major challenges facing us: where we are now, where we are heading in view of current trends; what practical steps could be taken to meet these challenges.
- Building a small Founding Council of respected individuals from among those who have been proposed for WFC membership, to advise and supervise the work of the WFC Initiative.
- Exploring, with international legal expertise, the best possible institutional structure for the WFC.
- Raising substantial funds to initiate the World Future Council.

11. WFC Launch and Operations

The WFC will only be launched when the basic funding for its initial phase has been guaranteed. Our aim is not to produce declarations and reports with no process for implementing them. We respect the work being done to develop good reform proposals and do not want to duplicate such work, but to build the framework for a process of implementation.

The initial WFC membership will be selected jointly by the WFC Founding Council and the WFC Initiative from individuals who have received the broadest support during the consultation process, prioritizing those who are respected internationally.

Concerning the legitimacy of this process, it is important to emphasize again that the WFC will ultimately be legitimized by the quality of its work. It will not claim to speak for others and will have no formal powers. All its proposals will have to be approved by democratically elected policy makers in order to become legally binding.

The process of consultation and selection will be the most broad-based and transparent ever held for any such body.

When this process has been completed, the WFC Initiative will hand over control to a WFC board, elected by its membership. This board will commission the WFC Charter, determine the procedures of membership rotation and the selection of new members. Once the WFC is incorporated in the country or countries in which it is head-quartered, it will be funded through national foundations, with trustees appointed by the WFC board. The WFC will meet at least annually to hold public hearings and make concrete recommendations, based on the research of its expert commissions.

12. Implementation

The WFC will be supplemented by advisory commissions on key issues needing global action. Currently we have identified 24 such issues, of which some are very broad and may require smaller sub-commissions. There is no need to duplicate existing work. We therefore propose that these commissions be hosted by recognized existing institutions, chosen by the WFC board. The WFC will have its own Research Department to liaise with and co-ordinate their work, to ensure that it fulfils the needs of the Council.

Cities who want to participate in the WFC, but cannot secure the funding required for its secretariats, will also be given the opportunity to host one or more of the commissions. Several institutions and cities have already expressed a strong interest in providing a 'home' for a WFC commission.

The WFC will support global change in a number of ways, through raising the level of public debate and providing a more inter-connected and values-based perspective. But the parliamentary connection is crucial if we are to bridge the current 'implementation gap', when even agreed global initiatives, like the Millennium Development Goals, are not implemented due to institutional blockages.

Reform proposals adopted by the WFC will be presented to MPs who have indicated an interest to work with the WFC in this area. Joint meetings will be held between them and WFC members to develop concrete recommendations (model legislation), to be put to the whole global network of democratic MPs. The e-Parliament Initiative is developing the facilities for such proposals to be voted on electronically. Proposals with majority support will then be introduced in national parliaments with the help of MPs. The WFC will also work to build worldwide support in national legislatures for the Council's recommendations.

The WFC will encourage the creation of national, regional and local future councils. These will be based on the Swiss model, already adopted in the Canton of Vaud, i.e. have the status of a constitutional body that decision-makers are obliged to take into account. Even without formal 'future veto' powers, the recommendations of such councils would have more weight than purely advisory commissions.

Significant reforms will naturally encounter opposition from vested interests and will not be achieved without substantial civil society support. One of the WFC founding partners, EarthAction, a global network of over 2000 Civil Society Organizations in 170 countries, has a successful track record in mobilizing such support through timely global campaigns and Action Alerts. EarthAction will help initiate and co-ordinate civil society networks on the various themes covered by the WFC commissions, to ensure focused and action-oriented civil society involvement in the WFC process. Through its global network it will engage organizations, citizens and the media to press policy makers to take action in support of the Council's proposals.

Jakob von Uexküll

Jakob von Uexküll, former member of the European Parliament, is the founder of the Right Livelihood Award in 1980. With the award he wants to recognize the efforts of those who are tackling issues more directly, coming up with practical answers to such challenges as the pollution of the environment, the abuse of basic human rights, the destitution and misery of the poor and the over-consumption and spiritual poverty of the wealthy.

Einstein's Dreams – a Scientific and Political Vision for our Future

Interview with Abhay Ashtekar conducted by Ekkehard Sieker[1]

What do you expect of the International Einstein Year in terms of science and politics, and especially regarding Einstein's ideas?

I feel that the world has unfortunately distanced itself considerably from Einstein's ideas. Here I'm referring to the great gap that has arisen between the Western countries and the Islamic world. We simply no longer think globally. The division is just too deep. Einstein, who was already a pacifist at the beginning of World War I, would be very sad to see what's happening in the world today.

I hope that at least some politicians – perhaps some wise politicians, such as Kofi Annan – will take Einstein's legacy to the peace movement seriously and will attempt to unite the world, so that we can pursue humanitarian goals together.

Perhaps it's too much to expect that Einstein's political dream of totally abolishing nuclear weapons, could actually become reality in the short term. It would be really wonderful if the countries that already possess nuclear weapons could get together with the countries developing them, and all these countries could sign the International Test Ban Treaty again, and could at least take a few initial steps towards the non-proliferation of nuclear matter.

It's a dream, but that's what I hope for, that during the International Einstein Year at least initial steps can be taken to enable people once again to hope and have their own visions, visions of long-term goals, and to get them to take a few first steps towards these goals. Of course, we won't be able to reach the final goals within this one year.

1) Ekkehard Sieker is a science journalist at the Max-Planck-Gesellschaft.

In addition to Einstein's political dream, did he have a scientific dream that can be clearly described?

Yes, of course! But to explain this dream clearly, I have to give you some background information in physics. At the beginning of the 20th century, we learned amazing things from quantum mechanics. Up until that point people thought that radiation and matter were two completely separate entities, but it was discovered that matter can actually be converted to radiation, such as light, and radiation can be converted to matter. In fact they are just two aspects of the same physical reality.

Then, in 1915, we learned from Einstein and his General Theory of Relativity that space-time geometry and matter fields have the same physical basis. In Einstein's field equations, geometry is actually present on one side of the equation – the space-time curvature – and on the other side are matter fields, and the two sides are equal. This raises the question of whether space-time geometry can be converted to matter. Without a quantum theory of space-time, this question can't be answered. Therefore Einstein wasn't able to answer it.

Then, in 1974, Stephen Hawking discovered that black holes actually emit radiation quantum mechanically by means of the tunnel effect. While they do this, they lose mass and shrink. This appears to be a transformation of geometry into radiation or matter. But this conclusion could not really be stated definitely, because Hawking's work was based on a space-time continuum – flat space-time; at that time he didn't have a quantum theory of space-time to work with.

Now we finally have the quantum theory of space-time geometry and can reconsider the phenomenon of Hawking radiation. It's literally true that the surface of a black hole is quantized in certain areas, which means that certain quantized areas exist on the surface of a black hole. When the black hole vaporizes through quantum effects, it literally shrinks. It loses one or two radiation quanta, and it really shrinks: in discrete steps, not continuously. While the black hole is shrinking and losing, say, one quantum of its space-time area, this space-time quantum is apparently converted into matter or radiation; it could be photons or electrons that emerge.

So it is very likely that there is a transformation of space-time quanta into matter and radiation. If this hypothesis is definitely proved, Einstein's scientific dream would have come true in the profoundest possible way.

What was your first contact with Einstein's ideas and Einstein himself?

The first real contact I had with Einstein's ideas came from a book written by a Russian-born American physicist, George Gamow, in which there is a theory of cosmology. So I read a little bit more about general relativity. And although I didn't understand too much, it seemed to me so miraculous. I was still in school at that time, and I was about 15 years old.

As Hermann Weyl said about general relativity: it is as though there's a veil through which we see reality, in a fuzzy way. And suddenly the veil disappears and vast expanses of truth and beauty become available to us. And that's what I felt, too, when I read these basic ideas of general relativity. As Francis Bacon would say: There was this strangeness in proportion. Strange things – but in exactly the right proportion. These were my first impressions which haven't changed at all, right up until today.

What could physics have in common with children's curiosity, with their way of asking questions?

I think everything! It's our responsibility to answer these questions. Sometimes it requires a lot of thinking because with these questions you have to put aside all equations and all the technical knowledge you've got in your head and go to the core of that question – and try to answer it.

My own son is six and a half and he asks these beautiful questions, which sometimes are really difficult. And I make every effort to answer them at the same level that he's asking them. That's not easy but it's what we all should be doing.

How would you explain to a kid, like my 8-year-old daughter, what you really do every day?

I would say: First of all, physics – the physical world around us – is much more subtle then we can perceive. Our perceptions can often be deceptive. They don't always capture the full story. For example, look at this table! This table looks to you like a completely smooth thing. But in fact it has atoms and it's not really completely continuous at all. It has this discrete structure. Then she would probably want to ask a lot of questions about atoms and such. After she had exhausted all her questions I would continue: Move your hands and you will feel the space around you. You will feel that the space is completely continuous: You can move your hand really slowly to any place and you don't jerk your hand, you can move it completely smoothly. And that's why we think – and have been taught for the last 1000 years – that space is continuous. But what we find is that even space has its tiny little atoms. Even space is not really continuous. And then I'd explain how these tiny atoms can get together and give an illusion of continuum. Finally I'd say: What I am working on is how to find properties, how these atoms of space – these one-dimensional 'threads' of geometry – which make up all this space, how these atoms behave, what their properties are.

I think this should be an ongoing dialog with your daughter and even with curious adults, not just in physics, but also in biology and genetics.

What do you wish would happen in the coming year, the International Einstein Year, both for scientists and the general public?

As far as the general public goes, I think that either we scientists are not doing our job right, or that because of society's views and values – probably because of both – fewer and fewer people seem to be interested in the fundamentals of science. Instead they spend time with things that require only a very short attention span, like the video-clips on the different music channels, and they are not accustomed to sustaining the longer attention spans that are essential to reaching a deeper understanding of nature.

From the point of view of the general public, it would be the best thing if people were actually able to comprehend these cherished ideas of Einstein's, namely the miracle of the existence of the universe and its many subtle properties. If more and more people actually feel drawn to science, not necessarily to becoming scientists – that would be a great thing, but not necessary – but to experiencing a feeling of awe and wonder for the miracles of nature, for the mysterious universe all around us, I think that would be very, very good. If, in addition to this, if more people were to find the courage to press for a more socially and economically just world without war, for a world in which people, especially people in the non-developed countries, had the opportunity to live in human dignity, to have enough food, medical care and of course a comprehensive education as well, then we would be pretty close to fulfilling Einstein's political dreams and social visions – which have lost none of their importance in today's world. If, during the International Einstein Year, we succeed in educating more people about Einstein's scientific and political ideas and giving them the courage to feel awe for nature, to remain curious and to ask open and critical questions about difficult scientific and social issues, this year will have been in the spirit of Einstein and will have fulfilled its purpose.

Abhay Ashtekar

Abhay Ashtekar was born in 1949 in India. He is Professor of Physics and the Director of the Center for Gravitational Physics and Geometry at the Pennsylvania State University, USA. He is one of the developers of the "Loop Quantum Gravitation Theory".

Part 6
Appendix

What Life Means to Einstein – An Interview by George Sylvester Viereck

(The Saturday Evening Post, October 26, 1929)

Relativity! What word is more symbolic of the age? We have ceased to be positive of anything. We look upon all things in the light of relativity. Relativity has become the plaything of the parlor philosopher. Is there any standard that has not been challenged in this our post-war world? Is there any absolute system of ethics, of economics or of law, whose stability or permanence is not assailed somewhere? Can there be any permanent value or absolute truth in a world in which the three angels of the triangle have ceased to be equal to two right angles – in a world in which time itself has lost its meaning; in which infinity becomes finite is lost in the infinite?

Einstein refuses to sponsor newfangled theories that draw their justification from his own assault upon the certainties of mathematics. His voice was bell-like and gentle, but his words were decisive when he smashed with one sentence the rash application of the term "relativity" to philosophy and to life. "The meaning of relativity", he said, "has been widely misunderstood. Philosophers play with the word, like a child with a doll. Relativity, as I see it, merely denotes that certain physical and mechanical facts, which have been regarded as positive and permanent, are relative with regard to certain other facts I the sphere of physics and mechanics. It does not mean that everything in life is relative and that we have a right to turn the whole world mischievously topsy-turvy." I now remember that some years ago, when I first met Einstein in New York, he had emphatically resisted the suggestion that he was a philosopher. "I am", he said, "solely a physicist." In spite of these denials, Einstein stands in a symbolic relation to our age – an age characterized by a revolt against the absolute in every sphere of science and of thought. He is a child of his age even if eschews metaphysic.

Einstein – Peace Now! Reiner Braun and David Krieger (Eds.)
Copyright © 2005 WILEY-VCH Verlag GmbH & Co. KGaA, Weinheim
ISBN 3-527-40604-2

A Born Teacher

Like Napoleon, like Mussolini, Albert Einstein has the distinction of having become an almost legendary figure in his own lifetime. No man since Copernicus, Galileo and Newton has wrought more fundamental changes in our attitude toward the universe. Einstein's universe is finite. Seen through Einstein's eyes, space and time are almost interchangeable terms. Time appears caparisoned as a fourth dimension. Space, once undefinable, has assumed the shape of the sphere. Einstein taught us that light travels in curves. All these facts are deducted from the theory of relativity advanced by Einstein in 1915.

With the advent of Einstein, mathematics ceased to be an exact science in the fashion of Euclid. The new mathematics appeared in the midst of the World War. It is not impossible that in the evolution of human thought Einstein's discovery may play a greater part than the Great War. His fame may outlive Foch and Ludendorff, Wilson and Clemenceau.

Einstein, in the words of his favourite colleague, Erwin Schrödinger, explains fundamental law of mechanics as geometrical proportions of space and time.

I shall not attempt to expound this statement. It is said that only ten men understand Einstein's theory of relativity.

Einstein's patience is infinite. He likes to explain his theories. A born teacher, Einstein does not resent questions. He loves children. The ten-year-old son of a friend was convinced that he had discovered the secret of perpetual motion. Einstein painstakingly explained to him the flaw in his calculations.

Whenever a question involving a difficult mathematical problem comes up, Einstein immediately takes up his pencil and covers page after page with the most intricate equations. He does not refer to a textbook; he works out such formulas immediately himself. Often the formula thus obtained is clearer, more comprehensible and more perfect than the equation that is found in books of reference.

Recently someone talked to him about color photography. Einstein immediately revolved the subject in his mind. He studied the camera, he made various calculations, and before the evening was over, he had evolved a new method of color photography.

It is difficult for him to explain his theories when he writes an article for lay consumption. But when the inquiring layman exposes the abysses of his ignorance face to face with Einstein, the great mathematician usually succeeds in bridging in gulf with an apt illustration. Talking to him, I saw in a flash not only a fourth dimension but numerous others. Glowing with pride in my achievement, I scribbled down a sentence here and there, but afterward my notes were as difficult to interpret as the fantastic network of dream.

"How can I form at least a dim idea of the fourth dimension?"

"Imagine," Einstein replied, slightly inclining his head with the crown of curly white hair," a scene in two-dimensional space – for instance, the painting of a man reclining on a bench. A tree stands beside the bench. Then imagine that a man walks from the bench to a rock on the other side of the tree. He cannot reach the rock except by walking either in front of or behind the tree. This is impossible in two-dimensional space. He can reach the rock only by an excursion into the third dimension.

"Now imagine another man sitting on the bench. How did he get there? Since two bodies cannot occupy the same place at the same time, he can have got only before or after the first man moved. He must have moved in time. Time is the fourth dimension. In a similar manner it is possible to explain five, six and more dimensions. Many problems of mathematics are simplified by assuming the existence of more dimensions."

I tried to secure an explanation of the fifth dimension. I regret to say that I do not remember the answer clearly. Einstein said something about a ball being thrown, which could disappear in one or two holes. One of these holes was a fifth, the other the sixth dimension.

I find it easier to understand Einstein's discovery, promulgated in 1929, which explains the universe in terms of electromagnetism. But, unfortunately, Einstein has not yet completely succeeded in convincing himself. He does not look upon the six pages that startled the world, pages immediately transmitted in facsimile across the ether, as a final conclusion.

To reach his conclusion, it was necessary for Einstein to express gravity in terms of electricity. The formula needed for this purpose is so complex, that in order to explain is meaning he was compelled to create a new system of advanced mathematics.

Einstein's new system reconciles Euclid with Riemann. It restores parallel lines, which Riemann abolished. According to Riemann, there can be no parallel lines in a curved universe. Einstein rediscovered parallel lines with the aid of the fourth dimension. Don't ask me to explain the process in detail. It is a thing that can be told in a series of intricate equations which no human being, not even Einstein himself, can visualize.

"No man," as Einstein said to me, sitting comfortably on the couch of the sitting room of his Berlin home, "can visualize four dimensions, except mathematically. We cannot visualize even three dimensions."

"But don't you," I said, "think in four dimensions?"

"I think in four dimensions," he replied, "but only abstractly. The human mind can picture these dimensions no more than it can envisage electricity. Nevertheless, they are no less real than electro-magnetism, the force which controls our universe, within and by which we have our being."

"I am particularly interested in your new theory which proves that gravity and electricity are one. Surly no six pages ever written by the hand of any scholar have so revolutionized human thought?"

"Unfortunately," Einstein remarked, with a smile, which gave a touch of impishness to his face, "my last theory is only a hypothesis which remain to be proved. It is different with my theory of relativity, which has been confirmed by many independent investigators and may now be regarded as definitely established."

Again a smile played about his face, creeping from his eyes toward his cheeks and disappearing in his moustache, slightly darker in color than the tangled mass of hair on his head.

Mrs. Einstein, his wife and cousin, as well as his helpmate, filled our glasses with strawberry juice and heaped more fruit salad upon our plates. Einstein never takes alcohol in any form, but he cannot resist the temptation of tobacco. He smokes more cigarettes than he should, with the guilty enjoyment of a schoolboy sporting his first cigar. It thrilled me to share strawberry juice and fruit salad with the man whose name is on ever lip and whose thoughts hardly anyone understands.

The close relationship between Einstein and his spouse express itself in the similarity of their foreheads. Their fathers were brothers and their mothers were sisters. "I am," Mrs. Einstein said quietly, "almost everything to my husband that it possible to be." Mrs. Einstein resembles a portrait of her sister, Mrs. Gumpertz, painted some years ago by Sir John Lavery, called The Lady with the Sables.

Einstein grew up with his cousin. They were friends from the very beginning. When fate separated them early in life, Einstein married a brilliant woman mathematician, a native of Serbia. Einstein has two children by his first wife. His childhood companion, the present Mrs. Einstein, also married and became the mother of a family. Her husband died after a few years of marriage. Then some force, stronger than those that Professor Einstein imprisons in his dynamic equations, drew the two cousins together. Albert Einstein secured a divorce from his mathematical wife and married his widowed cousin. Perhaps is a mistake for a physicist to marry a mathematician. There is James Huneker once remarked to me, no room in one family for two prima donnas.

The storm and stress of this period has graven its mark on Einstein's features and in his heart. Einstein's relations with his former wife are still friendly. He is deeply interested in the children of his first marriage, and he has adopted as his own the children sprung from his cousin's first union.

One of his commentators, Alexander Moskowski, calls Einstein a masculine sphinx. When Einstein speaks, his animated face reminds one somewhat of Briand, except that his features are more refined and more intellectual. If Briand espouses Pan-Europe, Einstein's vision embraces the world.

Einstein's struggles with fate have left no bitterness on his tongue. Every line of his face expresses kindliness. It also bespeaks indomitable pride. Some friends and admirers learned that he had decided to build a summer house with his hard-earned savings. They offered him a princely gift of land. But Einstein shook his head. "No," he said; "I could accept a gift from a community. I cannot accept such a gift from an individual. Every gift we accept is a tie. Sometimes," he added with Talmudic wisdom, "one pays most for the things one gets for nothing."

His Attic Retreat

Although the most-talked-about scientist of the world, Einstein absolutely refuses to capitalize his reputation. He laughed when he was asked to indorse an America cigarette. The money offered for his name would have paid the expense of his summer house. Knowing that fame has set him apart from other men, he feels that he must preserve at all cost the integrity of his soul. He escapes the interviewer by every possible device. His shyness dictates and his wife abets his seclusion. Unable to check the avalanche of offers and requests that overwhelm him, he leaves most letters, even from celebrities, unanswered. But he never ignores even the smallest note from a friend. He turned down princely offers to exploit his theories and his life in a book for popular consumption. "I refuse," he said again and again, "to make money out of my science. My laurel is not for sale like so many bales of cotton."

It is not generally known that Professor Einstein is not merely an expert in the upper regions of higher mathematics but that he takes a special delight in the practical solution of technical problems, such as confront the builder of machines and the electrician. His mind almost instinctively comes to conclusions that escape the ordinary engineer. He owes his training in his practical work to the fact that he was for several years an adviser to the Swiss paten office. It is through work of this type that Einstein has built up a modest fortune that enables him to build a house for himself without relying upon the munificence of the city of Berlin.

Einstein solves the mathematical and technical problems that are submitted to him in the solitude of his attic on the top floor of the apartment house in the Haberlandstrasse, where he lives. He furnished the little attic exclusively with the rather primitive furniture that he bought many years ago with his first savings.

I expected to see queer utensils and rare tomes in Einstein's secret retreat. I should not have been surprised if his den had resembled the laboratory of a medieval magician. I was doomed to disappointment. Einstein does not emulate Doctor Faust. There are a few books, also a few pictures. Faraday, Maxwell, Newton. I saw neither circles nor triangles. Einstein's only instrument is the head. He needs no books. His brain is his library.

From his desk Einstein sees only roofs – an ocean of roofs – and the sky. Here he is alone with his speculations. Here, Pallas-like, leaped from his head the theories that have revolutionized modern science. Here no human interference impedes the fight of his thoughts. Even his wife does no enter this holy of holies without trepidation.

Albert Einstein does not bury himself in his studies uninterruptedly. He is not a mollycoddle physically. He loves aquatic sports. His favourite toy is a sailboat with all modern technical improvements, in which he enjoys himself on the lakes and the rivers near his country place, Caputh. A towel wrapped fantastically around his head, he looks more like a pirate than like a professor of a great university. Battling with the wind, he forgets relativity and the fourth dimension. When the spray glistens in the silver of his hair and the sun strokes his cherub-like features, his thoughts are far from curved time–space.

Our Intellectual Democracy

A speculative thinker, a practical engineer, a sportsman and an artist, Einstein comes close to the Greek ideal of harmonious development. When he neither sales his boat not permits his mind to meander through fourth-dimensional space, Einstein enjoy himself with his violin. While I waited at the door of his apartment, it seemed to me that I heard strains of elfin music. Perhaps it was Einstein playing. When I entered, he was wrapping up his violin for the night like a mother putting her child to bed.

Professor Einstein looks more like a musician than like a mathematician. "If," he confessed to me, with a smile that was half-wistful, half-apologetic, "I were not a physicist, I would probably be a musician. I often think in music. I live my daydreams in music. I see my life in terms of music."

"Perhaps," I remarked, "if you had chosen to become a musician you would outshine Richard Strauss and Schönberg. Perhaps you would have given us the music of the spheres or a fourth-dimensional music."

Einstein gazed dreamily – was it into the far corners of the room, or was it into space – that space that his investigations have robbed of infinity? "I cannot tell," he replied, "if I would have done any creative work of importance in music, but I do know that I get most joy in life out of my violin." As a matter of fact, Einstein's taste in music is severely classical. Even Wagner is to him no unalloyed feast of the ears. He adores Mozart and Bach. He even prefers their work to the architectural music of Beethoven.

President Hindenburg hardly ever appears in public, because he is immediately recognized wherever he goes. For the same reason, Professor Einstein refuses all invitations to the more popular restaurants. Although his world fame compels him to seek isolation, he is a sociable being. He loves quiet chats over his own dinner table with such friends as Gerhart Hauptmann and Professor Schrödinger. He reads only little. Modern fiction does not seduce him. Even in science he limits himself largely to his special field. "Reading after a certain age diverts the mind too much from its creative pursuits. Any man who reads too much and uses his own brain too little falls into lazy habits of thinking, just as the man who spends too much time in the theatre is tempted to be content with living vicariously instead of living his own life."

In his own field of thought Einstein follows every development with keen interest. He has a gift of reading at a glance a whole page of equations. Einstein can master a whole new system of mathematics in half an hour.

"Who," I asked to him, "are you greatest contemporaries?"

"I cannot reply to this question," Einstein answered, his eyes twinkling humorously, "without compiling an encyclopaedia. I cannot even discuss intelligently the men who labour in my own field without writing a book."

"Our time," he added, "is Gothic in its spirit. Unlike the Renaissance, it is not dominated by a few outstanding personalities. The twentieth century has established the democracy of intellect. In the republic of art and science there are many men who take an equally important part in the intellectual movements of our age. It is epoch rather than the individual that is important. There is no one dominant personality like Galileo or Newton. Even I the nineteenth century there were still a few giants who out topped all others. Today the general level is much higher than ever before in the history of the world, but there are few men whose stature immediately sets them apart from all others."

"Whom do you consider the most conspicuous worker in your own field?"

The Contemporary Great

"It is not fair," Einstein replied, "to single out individuals. In Germany, I consider Schrödinger and Heisenberg as being of special importance."

"Schrödinger?" I said. "What has he done?"

"Schrödinger has discovered the mathematics formula of the fact that all life moves in waves."

"And Heisenberg?"

"Heisenberg is a sovereign mathematician who has formulated a new definition of mathematical magnitudes. Then there is, of course, Planck, the exponent of the quantum theory."

I did not ask Einstein to explain the quantum theory. I know that it is even more difficult to grasp than relativity.

"Would you say that Eddington is your most brilliant interpreter?"

"Eddington," Einstein replied, "is a great mathematician, but his supreme achievement is his discovery of the physical constitution of the stars."

"Is there," I asked modestly, "anyone in America whose importance is commensurable with that of the men you have just discussed?"

"In America," Einstein replied quietly, "more than anywhere else, the individual is lost in the achievements of the many. America is beginning to be the world leader in scientific investigation. America scholarship is both patient and inspiring. The Americans show an unselfish devotion to science, which is the very opposite of the conventional European view of your countrymen. Too many of us look upon Americans as dollar chasers. This is a cruel libel, even if it is reiterated thoughtlessly by the Americans themselves. It is not true that the dollar is an America fetish. The American students are not interested in dollars, not even in success as such, but in his task, the object of the search. It is his painstaking application to the study of the infinitely little and the infinitely large which accounts for his success in astronomy."

"What," I asked, "have been our most outstanding accomplishments in your field?"

"America," Einstein replied, "has been especially successful in increasing our knowledge in the fixed stars. But in Holland and elsewhere there are many who have done remarkable work."

"The Americans," Einstein continued, "are idealists. Wilson, notwithstanding the collapse of his Fourteen Points, was inspired by high ideals. America entered the war for idealistic reasons, in spite of the fact that material interests were exerting the utmost pressure in the same direction."

"We are inclined" – Einstein inclined his head lightly to one side like a bird – "to overemphasize the material influence in history. The Russian especially make this mistake. Intellectual values and ethnic influences, tradition and emotional factors are equally important. If this were not the case, Europe would be today a federated state, not a madhouse of nationalism."

Born in Ulm, in Germany, in 1879, educated partly there, partly in Italy and partly in Switzerland, a Swiss as well as a German citizen, Einstein regards international jealousies with the serenity with which a teacher looks upon quarrelling schoolboys. In politics he leans to Socialism. He looks upon pacifism as the ultimate ideal. Poor, a Jew, a Socialist and a pacifist, Einstein carried four handicaps like millstones around the neck. Einstein conquers all obstacles, including his own shyness, by the sheer force of his celebration. He does not reject any form of government except absolutism. He is tolerant, but by no means uncritical, in his attitude toward Russia.

"What," I inquired, "is your attitude toward Bolshevism?"

"Bolshevism is an extraordinary experiment. It is not impossible that the drift of social evolution henceforward may be in the direction of communism. The Bolshevist experiment may be worth trying. But I think that Russia errs badly in the execution of her ideal. The Russians makes mistake of putting party faith above efficiency. They replace efficient men by politicians. Their test stone of public service is not the accomplishment but devotion to a rigid creed."

"Do you believe in the German Republic?"

"Undoubtedly. The people have the right to rule themselves. Now, at least, our mistakes are our own."

We Can Do What We Wish, But –

"Do you blame the Kaiser for the downfall of Germany?"

"The Kaiser," Einstein replied, "meant well. He often had the right instincts. His intuitions were frequently more inspired that the laboured reasons of his Foreign Office. Unfortunately, the Kaiser was always surrounded by poor advisers."

"It seems to me," I interjected, "that there are two parties in Germany. One blames Kaiser for the German debacle, the other attempts to saddle the responsibility upon the Jews."

"Both," Einstein remarked, "are largely guiltless. The German debacle was due to the fact that the German people, especially the upper classes, failed to produce men of character, strong enough to take hold of the reins of government and to tell the truth to the Kaiser.

"It was partly," Einstein added somewhat hesitatingly, "the fault of Bismarck. Bismarck's philosophy of government was wrong. Besides, there was no one to succeed to the giant. Like many men of genius, he was too jealous to permit any other man to walk in his footsteps. In fact, it is doubtful if any other man could have followed the tortuous path of Bismarckian politics."

"In a sense," he added, "we can hold no one responsible. I am a determinist. As such, I do not believe in free will. They believe that man shapes his own life. I reject that doctrine philosophically. In that respect I am not a Jew."

"Don't you believe that man is a free agent at least in a limited sense?"

Einstein smiled ingratiatingly. "I believe with Schopenhauer: We can do what we wish, but we can only wish what we must. Practically, I am, nevertheless, compelled to act as if freedom of the will existed. If I wish to live in a civilized community, I must act as if man is a responsible being."

"I know that philosophically a murderer is not responsible for his crime; nevertheless, I must protect myself from unpleasant contacts. I may consider him guiltless, but I prefer not to take tea with him."

"Do you mean to say that you did not choose your own career, but that your actions were predetermined by some power outside of yourself?"

The Danger of Too Much Analysis

"My own career was undoubtedly determined, not by my own will but by various factors over which I have no control – primarily those mysterious glands in which Nature prepares the very essence of life, our internal secretions."

"It may interest you," I interjected, "that Henry Ford once told me that he, too, did not carve out his own life, but that all his actions were determined by an inner voice."

"Ford," Einstein replied, "may call it his inner voice. Socrates referred to it as his daimon. We moderns prefer to speak of our glands of internal secretion. Each explains in his own way the undeniable fact that the human will is not free."

"Don't you deliberately ignore all physic factors in human development? What, for instance, I asked "is your attitude toward the subconscious? According to Freud, psychic events registered indelibly in our nether mind make and mar our lives."

"Whereas materialistic historians and philosophers neglect psychic realities, Freud is inclined to overstress their importance. I am not a psychologist, but it seems to me fairly evident that physiological factors, especially our endocrines, control our destiny."

"Then you do not believe in psychoanalysis?"

"I am not," Einstein modestly replied, "able to venture a judgement on so important a phase of modern thought. However, it seems to me that psychoanalysis is not always salutary. It may not always be helpful to delve into the subconscious. The machinery of our legs is controlled by a hundred different muscles. Do you think it would helps us to walk if analyzed our legs and knew exactly which one of the little muscles must be employed in locomotion and the order in which they work?"

"Perhaps," he added with the whimsical smile that sometimes lights up the somber pools of his eyes like a will-o'-the-wisp, "you remember the story of the toad and the centipede? The centipede was very proud of having one hundred legs. His neighbour, the toad, was very much depressed because he had only four. One day a diabolic inspiration prompted the toad to write a letter to the centipede as follows:"

"Honoured Sir: Can you tell me which one of your hundred legs you move first, when you transfer your distinguished body from one place to another, and in what order you move the other ninety-nine legs?"
"When the centipede received this letter he began to think. He tried first one leg, then the other. Finally he discovered to his consternation that he was unable to move a single leg. He could no longer walk at all! He was paralyzed! It is possible that analysis may paralyze our mental and emotional processes in a similar manner."

"Are you then an opponent of Freud?"

"By no means. I am not prepared to accept all his conclusions, but I am consider his work an immensely valuable contribution to the science of human behavior. I think he is even greater as a writer than as a psychologist. Freud's brilliant style is unsurpassed by anyone since Schopenhauer."

There was a pause, filled by more fruit salad and strawberry juice.

"Is there", I resumed the conversation, "such a thing as progress in the story of human effort?"

"The only progress I can see is progress in organization. The ordinary human being does not live long enough to draw any substantial benefit from his own experience. And no one, it seems, can benefit by the experience of others. Being both a father and a teacher. I know we can teach our children nothing. We can transmit to them neither our knowledge of life nor of mathematics. Each must learn its lesson anew."

"But", I interjected, "nature crystallizes our experiences. the experiences of one generation are the instincts of the next."

"Ah," Einstein remarked, "that is true. But it takes Nature ten thousand or ten million of years to transmit inherited experience of characteristics. It must have taken the bees and the ants aeons before they learned to adapt themselves so marvellously to their environments. Human beings, alas, seem to learn more slowly than insects."

"Do you think that mankind will eventually evolve the superman?"

"If so," Einstein replied, "it will be a matter of millions of years."

"You don't agree with Nitzesche's sister that Mussolini is the superman prophesied by her brother?"

Again a smile illuminated Einstein's features, but it was not so jovial as before. A pacifist and internationalist, Einstein is the very antithesis of the dictator.

Although he denies a freedom of the will philosophically, Einstein resents any attempt to circumscribe still further the limited sphere within which the human will may exert itself with the illusion of freedom.

"If we owe so little to the experience of others, how do you account for sudden leaps forward in the sphere of science? Do you ascribe your own discoveries to intuition or inspiration?"

The Measles of Mankind

"I believe in intuitions and inspirations. I sometimes feel that I am right. I do not know that I am. When two expeditions of scientists, financed by the Royal Academy, went forth to test my theory of relativity, I was convinced that their conclusions would tally with my hypothesis. I was not surprised when the eclipse of May 29, 1919, confirmed my intuitions. I would have been surprised if I had been wrong."

"Then you trust more to your imagination than to your knowledge?"

"I am enough of the artist to draw freely upon my imagination. Imagination is more important than knowledge. Knowledge is limited. Imagination encircles the world."

"To what extent are you influenced by Christianity?"

"As a child, I received instruction both in the Bible and in the Talmud. I am a Jew, but I am enthralled by the luminous figure of the Nazarene."

Have you read Emil Ludwig's book on Jesus?"

"Emil Ludwig's Jesus," Einstein replied, "is shallow. Jesus is too colossal for the pen of phrasemongers, however artful. No man can dispose of Christianity with a bon mot."

"You accept the historical existence of Jesus?"

"Unquestionably. No one can read the Gospels without feeling the actual presence of Jesus. His personality pulsates in every word. No myth is filled with such life. How different, for instance, is the impression which we receive from an account of legendary heroes of antiquity like Theseus. Theseus and another heroes of his type lack the authentic vitality of Jesus."

"Ludwig Lewisohn, in one of his recent books, claims that many of the sayings of Jesus paraphrase the sayings of other prophets."

"No man," Einstein replied, "can deny the fact that Jesus existed, nor that his sayings are beautiful. Even if some of them have been said before, no one has expressed them so divinely as he."

"Gilbert Chersterton told me that, according to a Catholic writer in a Dublin Review, your theory of relativity merely confirms the cosmology of Thomas Aquinas."

"I have not," Einstein replied, "read all the works of Thomas Aquinas, but I am delighted if I have reached the same conclusions as the comprehensive mind of that great Catholic scholar."

"Do you look upon yourself as a German or as a Jew?"

"It is quite possible," Einstein replied, "to be both. I look upon myself as man. Nationalism is an infantile disease. It is the measles of mankind."

The Standardization Peril

"How then," I said, "do you justify Jewish nationalism?"

"I support Zionism," Professor Einstein replied, "in spite of the fact that it is a national experiment, because it gives us Jews a common interest. This nationalism is no menace to other peoples. Zion is too small to develop the imperialistic designs."

"Then you do not believe in assimilation?"

"We Jews," Einstein replied, "have been too adaptable. We have been too eager to sacrifice our idiosyncrasies for the sake of social conformity."

"Perhaps assimilation makes for greater happiness."

"I don't think so," Einstein replied. "Even in modern civilization, the Jew is most happy if he remains a Jew."

"Do you believe in race as a substitute for nationalism?"

"Race, at least, constitutes a larger unit. Nevertheless, I do not believe in race as such. Race is fraud. All modern people are a conglomeration of so many ethnic mixtures that no pure race remains."

"Do you," I remarked, "look upon religion as the tie which holds the children of Israel together?"

"I do not think," Einstein replied thoughtfully, "that religion is the most important element. We are held together rather by a body of tradition, handed down from father to son, which the child imbibes with his mother's milk. The atmosphere of our infancy predetermines our idiosyncrasies and predilections. When I met you, I knew I could talk to you freely without the inhibitions which make the contact with others so difficult. I looked upon you not as German nor as an American but as a Jew."

"I have written the autobiography of the Wandering Jew with Paul Eldridge," I told him. "Nevertheless, it so happens that I am not a Jew. My parents and my progenitors are Nordics from Protestant Germany."

"It is possible," Professor Einstein observed, "for any individual to trace every drop of blood in his constitution. Ancestors multiply like the famous seed of corn on the chessboard which embarrassed the sultan. After we go back a few generations, our ancestors increase so prodigiously that it is practically impossible to determine exactly the various elements which constitute our being. You have the psychic adaptability of the Jew. There is something in your psychology which makes it possible for me to talk to you without barrier."

"Why should quickness of mind be only a Jewish characteristic? It is not also possessed by the Irish and to a large extent by the Americans?"

"Americans undoubtedly owe much to the melting pot. It is possible that this mixture of races makes their nationalism less objectionable than the nationalism of Europe. This may be due partly to the fact that your country is so immense, that you do not think in terms of narrow borders. It may be due to the fact that you do not suffer from the heritage of hatred or fear which poisons the relations of the nations of Europe."
"But to return to the Jewish questions. Other groups and nations cultivate their individual traditions. There is no reason why we should sacrifice ours. Standardization robs life of its spice. To deprive every ethnic group of its special traditions is to convert the world into a huge Ford plant. I believe in standardizing automobiles. I do not believe in standardizing human beings. Standardization is a great peril which threatens American culture."

"Do you consider Ford, then, a menace?"

"Ford is undoubtedly a man of genius. No man can create what Ford has created, unless the life force has provided him with conspicuous gifts. Nevertheless, I am sometimes sorry for men like Ford. Everybody who comes to them wants something form them. Such men do not always realize that the adoration which they receive is not a tribute to their personality but to their power and to their pocketbook. Great captains of industry and great kings fall into the same error. An invisible wall impedes their vision.

"I am happy because I want nothing from anyone. I do not care for money. Decorations, titles or distinctions mean nothing to me. I do not crave praise. The only thing that gives me pleasure, apart from my work, my violin and my sailboat, is the appreciation of my fellow workers."

"Your modesty," I remarked, "does you credit."

"No," Einstein replied with a shrug of his shoulders. "I claim credit for nothing. Everything is determined, the beginning as well as the end, by forces over which we have no control. It is determined for the insect as well as for the star. Human beings, vegetables or cosmic dust, we all dance to a mysterious tune, intoned in the distance by an invisible player."

Mrs. Einstein on Guard

Einstein rose and excused himself. It was nearly midnight. We had been talking for nearly three hours.

"My husband," Mrs. Einstein remarked, "must attend to important work. But there is no reason why you should go. Will you not stay here and talk to me?"

We talked and talked.

A little while later I saw the figure if Einstein, wrapped in a bathrobe, on his way to his daily ablution.

He smiled at me with the same droll smile that had captivated me from the beginning. It is something to have seen the sage in his bathrobe! The touch of common humanity in no way detracted from his dignity.

Mrs. Einstein´s eyes followed her husband adoringly when he vanished, and again when he reappeared from his bath. She adjusts herself to her husband in a tact that is rare in wives of great men.

When he ascends to his attic, she does not cling to his coat tails. When he wishes to be alone, she completely eliminates herself from his life. She spares him disharmonious contacts and protects the serenity of his mind with the devotion of a vestal virgin guarding the scared fire. It is by no means impossible that with a less-sacrificing mate, Einstein would not have made the discoveries that link his name with the immortals. Thus love, that moves the sun and all the stars, sustains in its lonely path the genius of Albert Einstein.

Appeal to the Europeans

Mid-October 1914

Never before has any war so completely disrupted cultural cooperation. It has done so at the very time when progress in technology and
communications clearly suggest that we recognize the need for international relations that will necessarily move in the direction of a
universal, world-wide civilization. Perhaps we are all the more keenly and painfully aware of the rupture precisely because so many international bonds existed before.

We can scarcely be surprised. Anyone who cares in the least for a
common world culture is now doubly committed to fight for the
maintenance of the principles on which it must stand. Yet, those
from whom such sentiments might have been expected – primarily
scientists and artists – have so far responded, almost to a man, as
though they had relinquished any further desire for the continuance
of international relations. They have spoken in a hostile spirit, and
they have failed to speak out for peace.

Nationalist passions cannot excuse this attitude that is unworthy
of what the world has heretofore called culture. It would be a grave
misfortune were this spirit to gain general currency among the intellectuals. It would, we are convinced, not only threaten culture as
such; it would endanger the very existence of the nations for the protection of which this barbarous war was unleashed.

Technology has shrunk the world. Indeed, today the nations of the
great European peninsula seem to jostle one another much as once
did the city-states that were crowded into those smaller peninsulas
jutting out into the Mediterranean. Travel is so widespread, international supply and demand are so interwoven, that Europe – one could
almost say the whole world – is even now a single unit.

Surely, it is the duty of the Europeans of education and good will
at least to try to prevent Europe from succumbing, because of lack of
international organization, to the fate that once engulfed ancient

Einstein – Peace Now! Reiner Braun and David Krieger (Eds.)

Greece! Or will Europe also suffer slow exhaustion and death by fratricidal war?

The struggle raging today can scarcely yield a "victor"; all nations that participate in it will, in all likelihood, pay an exceedingly high price. Hence it appears not only wise but imperative or men of education in all countries to exert their influence for the kind of peace treaty that will not carry the seeds of future wars, whatever the outcome of the present conflict may be. The unstable and fluid situation in Europe, created by the war, must be utilized to weld the Continent into an organic whole. Technically and intellectually, conditions are ripe for such a development.

This is not the place to discuss how this new order in Europe may be brought about. Our sole purpose is to affirm our profound conviction that the time has come when Europe must unite to guard its soil, its people, and its culture. We are stating publicly our faith in European unity, a faith that we believe is shared by many; we hope that this public affirmation of our faith may contribute to the growth of a powerful movement toward such unity.

The first step in this direction would be for all those who truly cherish the culture of Europe to join forces – all those whom Goethe once prophetically called "good Europeans." We must not abandon hope that their voice speaking in unison may even today rise above the clash of arms, particularly if they are joined by those who already enjoy renown and authority.

The first step, we repeat, is for Europeans to join forces. If, as we devoutly hope, enough Europeans are to be found in Europe – people to whom Europe is a vital cause rather than a geographical term – we shall endeavor to organize a League of Europeans. This league may then raise its voice and take action.

We ourselves seek only to make the first move, to issue the challenge. If you are of one mind with us, if you too are determined to create a widespread movement for European unity, we bid you pledge yourself by signing your name.

The Russell – Einstein Manifesto

Issued in London, 9 July 1955

In the tragic situation which confronts humanity, we feel that scientists should assemble in conference to appraise the perils that have arisen as a result of the development of weapons of mass destruction, and to discuss a resolution in the spirit of the appended draft.

We are speaking on this occasion, not as members of this or that nation, continent, or creed, but as human beings, members of the species Man, whose continued existence is in doubt. The world is full of conflicts; and, overshadowing all minor conflicts, the titanic struggle between Communism and anti-Communism.

Almost everybody who is politically conscious has strong feelings about one or more of these issues; but we want you, if you can, to set aside such feelings and consider yourselves only as members of a biological species which has had a remarkable history, and whose disappearance none of us can desire.

We shall try to say no single word which should appeal to one group rather than to another. All, equally, are in peril, and, if the peril is understood, there is hope that they may collectively avert it.

We have to learn to think in a new way. We have to learn to ask ourselves, not what steps can be taken to give military victory to whatever group we prefer, for there no longer are such steps; the question we have to ask ourselves is: what steps can be taken to prevent a military contest of which the issue must be disastrous to all parties?

The general public, and even many men in positions of authority, have not realized what would be involved in a war with nuclear bombs. The general public still thinks in terms of the obliteration of cities. It is understood that the new bombs are more powerful than the old, and that, while one A-bomb could obliterate Hiroshima, one H-bomb could obliterate the largest cities, such as London, New York, and Moscow.

Einstein – Peace Now! Reiner Braun and David Krieger (Eds.)
Copyright © 2005 WILEY-VCH Verlag GmbH & Co. KGaA, Weinheim
ISBN 3-527-40604-2

No doubt in an H-bomb war great cities would be obliterated. But this is one of the minor disasters that would have to be faced. If everybody in London, New York, and Moscow were exterminated, the world might, in the course of a few centuries, recover from the blow. But we now know, especially since the Bikini test, that nuclear bombs can gradually spread destruction over a very much wider area than had been supposed.

It is stated on very good authority that a bomb can now be manufactured which will be 2,500 times as powerful as that which destroyed Hiroshima. Such a bomb, if exploded near the ground or under water, sends radio-active particles into the upper air. They sink gradually and reach the surface of the earth in the form of a deadly dust or rain. It was this dust which infected the Japanese fishermen and their catch of fish.

No one knows how widely such lethal radio-active particles might be diffused, but the best authorities are unanimous in saying that a war with H-bombs might possibly put an end to the human race. It is feared that if many H-bombs are used there will be universal death, sudden only for a minority, but for the majority a slow torture of disease and disintegration.

Many warnings have been uttered by eminent men of science and by authorities in military strategy. None of them will say that the worst results are certain. What they do say is that these results are possible, and no one can be sure that they will not be realized. We have not yet found that the views of experts on this question depend in any degree upon their politics or prejudices. They depend only, so far as our researches have revealed, upon the extent of the particular expert's knowledge. We have found that the men who know most are the most gloomy.

Here, then, is the problem which we present to you, stark and dreadful and inescapable: Shall we put an end to the human race; or shall mankind renounce war? People will not face this alternative because it is so difficult to abolish war.

The abolition of war will demand distasteful limitations of national sovereignty. But what perhaps impedes understanding of the situation more than anything else is that the term "mankind" feels vague and abstract. People scarcely realize in imagination that the danger is to themselves and their children and their grandchildren, and not only to a dimly apprehended humanity. They can scarcely bring themselves to grasp that they, individually, and those whom they love are in imminent danger of perishing agonizingly. And so they hope that perhaps war may be allowed to continue provided modern weapons are prohibited.

This hope is illusory. Whatever agreements not to use H-bombs had been reached in time of peace, they would no longer be considered binding in time of war, and both sides would set to work to manufacture H-bombs as soon as war broke out, for, if one side manufactured the bombs and the other did not, the side that manufactured them would inevitably be victorious.

Although an agreement to renounce nuclear weapons as part of a general reduction of armaments would not afford an ultimate solution, it would serve certain important purposes. First: any agreement between East and West is to the good in so far as it tends to diminish tension. Second: the abolition of thermo-nuclear weapons, if each side believed that the other had carried it out sincerely, would lessen the fear of a sudden attack in the style of Pearl Harbour, which at present keeps both sides in a state of nervous apprehension. We should, therefore, welcome such an agreement though only as a first step.

Most of us are not neutral in feeling, but, as human beings, we have to remember that, if the issues between East and West are to be decided in any manner that can give any possible satisfaction to anybody, whether Communist or anti-Communist, whether Asian or European or American, whether White or Black, then these issues must not be decided by war. We should wish this to be understood, both in the East and in the West.

There lies before us, if we choose, continual progress in happiness, knowledge, and wisdom. Shall we, instead, choose death, because we cannot forget our quarrels? We appeal, as human beings, to human beings: Remember your humanity, and forget the rest. If you can do so, the way lies open to a new Paradise; if you cannot, there lies before you the risk of universal death.

Resolution

We invite this Congress, and through it the scientists of
the world and the general public, to subscribe to
the following resolution:

„In view of the fact that in any future world war nuclear weapons will certainly be employed, and that such weapons threaten the continued existence of mankind, we urge the Governments of the world to realize, and to acknowledge publicly, that their purpose cannot be furthered by a world war, and we urge them, consequently, to find peaceful means for the settlement of all matters of dispute between them."

Max Born

Perry W. Bridgman

Albert Einstein

Leopold Infeld

Frédéric Joliot-Curie

Herman J. Muller

Linus Pauling

Cecil F. Powell

Joseph Rotblat

Bertrand Russell

Hideki Yukawa

An Appeal to Stop the Spread of Nuclear Weapons

Ava Helen Pauling and Linus Pauling
3500 Fairpoint St., Pasadena, California

To the United Nations and to all nations in the world:

We, the men and women whose names are signed below, believe that stockpiles of nuclear weapons should not be allowed to spread to any more nations or groups of nations.

The world is now in great danger. A cataclysmic nuclear war might reach out as the result of some terrible accident or of an explosive deterioration in international relations such that even the wisest leaders would be unable to avert the catastrophe. Universal disarmament has now become the essential basis for life and liberty for all people.

The difficult problem of achieving universal disarmament would become far more difficult if more nations or groups of nations were to come into possession of nuclear weapons. We accordingly urge that the present nuclear powers not transfer nuclear weapons to other nations or groups of nations such as the North Atlantic Treaty Organisation or the Warsaw Pact group, that all nations not now possessing these weapons voluntarily refrain from obtaining or developing them, and that the United Nations and all nations increase their efforts to achieve total and universal disarmament with a system of international controls and inspection such as to insure to the greatest possible extent the safety of all nations and all people.

Linus Pauling
Ava Helen Pauling

Einstein – Peace Now! Reiner Braun and David Krieger (Eds.)
Copyright © 2005 WILEY-VCH Verlag GmbH & Co. KGaA, Weinheim
ISBN 3-527-40604-2

Appeal – End the Nuclear Weapons Threat to Humanity!

Nuclear Age Peace Foundation

We cannot hide from the threat that nuclear weapons pose to humanity and all life. These are not ordinary weapons, but instruments of mass annihilation that could destroy civilization and end most of life on Earth.

Nuclear weapons are morally and legally unjustifiable. They destroy indiscriminately – soldiers and civilians; men, women and children; the aged and the newly born; the healthy and the infirm.

The obligation to achieve nuclear disarmament "in all its aspects," as unanimously affirmed by the International Court of Justice, is at the heart of the Non-Proliferation Treaty.

More than ten years have now passed since the end of the Cold War, and yet nuclear weapons continue to cloud humanity's future. The only way to assure that nuclear weapons will not be used again is to abolish them.

We, therefore, call upon the leaders of the nations of the world and, in particular, the leaders of the nuclear weapons states to act now for the benefit of all humanity by taking the following steps:

- De-alert all nuclear weapons and de-couple all nuclear warheads from their delivery vehicles.
- Ratify the Comprehensive Test Ban Treaty.
- Commence good-faith negotiations to achieve a Nuclear Weapons Convention requiring the phased elimination of all nuclear weapons, with provisions for effective verification and enforcement.
- Declare policies of No First Use of nuclear weapons against other nuclear weapons states and policies of No Use against non-nuclear weapons states.
- Reallocate resources from the tens of billions of dollars currently being spent for maintaining nuclear arsenals to improving human health, education and welfare throughout the world.

Einstein – Peace Now! Reiner Braun and David Krieger (Eds.)
Copyright © 2005 WILEY-VCH Verlag GmbH & Co. KGaA, Weinheim
ISBN 3-527-40604-2

Hafsat Abiola, Mayor Tadatoshi Akiba (Hiroshima), Muhammad Ali, Isabel Allende, Oscar Arias**, Kenneth J. Arrow*, Armand Assante, Lloyd Axworthy, Rev. J. Edwin Bacon, Jr., Nicolaas Bloembergen*, Julian Bond, Elisabeth Mann Borgese, Howard Brembeck, Ambassador Richard Butler, Rev. Joan Brown Campbell, Rodrigo Carazo Odio, Jimmy Carter**, Admiral Eugene J. Carroll, Jr., Lenedra Carroll, Yvon Chouinard, Jean-Michel Cousteau, Alan Cranston, Walter Cronkite, Paul Crutzen*, The XIVth Dalai Lama**, Diandra Douglas, Michael Douglas, Marian Wright Edelman, Paul Erhlich, Richard R. Ernst*, Adolfo Perez Esquivel**, Richard Falk, Edmond H. Fischer*, Harrison Ford, John Kenneth Galbraith, Johan Galtung, Arun Gandhi, Admiral Noel Gayler, Donald A. Glaser*, Jane Goodall, Mikhail Gorbachev**, Nadine Gordimer*, Jonathan Granoff, Corbin Harney, Minoru Hataguchi, Alan J. Heeger*, Rev. Theodore M. Hesburgh, David H. Hubel*, Daisaku Ikeda, Mayor Iccho Itoh (Nagasaki), Craig Kielburger, Coretta Scott King, Lawrence R. Klein*, F.W. de Klerk**, Walter Kohn*, David Krieger, Dennis J. Kucinich, Admiral Gene R. La Rocque, Ambassador James Leonard, Sally Lilienthal, Bernard Lown, M.D.**, Wangari Maathai, Mairead Corrigan Maguire**, Peter Matthiessen, Rigoberta Menchú Tum**, Franco Modigliani*, Rev. James Parks Morton, Robert Muller, Kary B. Mullis*, Joseph E. Murray, M.D.*, Erwin Neher*, Paul Newman, Queen Noor of Jordan, Kenzaburo Oe*, John C. Polanyi*, Admiral L. Ramdas, Jose Ramos-Horta**, Rev. George F. Regas, Frederick C. Robbins*, Richard J. Roberts*, Senator Douglas Roche, Sir Joseph Rotblat**, Frederick Sanger*, Martin Sheen, Stanley K. Sheinbaum, Carly Simon, Jennifer Allen Simons, Michael Smith*, Gerry Spence, Jack Steinberger*, Meryl Streep, Barbra Streisand, Maj Britt Theorin**, E. Donnall Thomas*, Ted Turner, Archbishop Desmond Tutu**, Mordechai Vanunu, Ambassador Paul C. Warnke, Elie Wiesel**, Betty Williams**, Jody Williams**, Terry Tempest Williams, Joanne Woodward, Alla Yaroshinskaya

* *Nobel Laureate*
** *Nobel Peace Laureate*

01/3/2003

U.S.-Nobel-Laureates Object to Preventive Attack on Iraq

U.S. security and standing in the world would be undermined

Forty-one American Nobel laureates in science and economics issued a declaration yesterday opposing a preventive war against Iraq without wide international support. The statement, four sentences long, argues that an American attack would ultimately hurt the security and standing of the United States, even if it succeeds.

Military operations without a mandate of the United Nations could indeed lead to a relatively swift victory in the short run, but the longterm consequences would be very damaging for the US. "[...] the medical, economic, environmental, moral, spiritual, political, and legal consequences of an American preventive attack on Iraq would undermine, not protect, U.S. security and standing in the world", reads the 74-words declaration. Among the signatories is Hans Bethe, the physics laureate of 1967, who as member of the Manhattan Project took part in the developing of the first atomic bomb.

The signers are these, with E designating economics; P, physics; C, chemistry; and M, medicine or physiology:

George A. Akerlof E	Roger Guillemin M
Philip W. Anderson P	Herbert A. Hauptman C
Paul Berg C	Alan J. Heeger C
Hans A. Bethe P	Louis J. Ignarro M
Nicolaas Bloembergen P	Eric R. Kandel M
Paul D. Boyer C	Har Gobind Khorana M
Owen Chamberlain P	Lawrence R. Klein E
Leon N. Cooper P	Walter Kohn C
James W. Cronin P	Leon M. Lederman P
Robert F. Curl Jr. C	Yuan T. Lee C
Val L. Fitch P	William N. Lipscomb C
Robert F. Furchgott M	Daniel L. McFadden E
Sheldon L. Glashow P	Franco Modigliani E

Einstein – Peace Now! Reiner Braun and David Krieger (Eds.)
Copyright © 2005 WILEY-VCH Verlag GmbH & Co. KGaA, Weinheim
ISBN 3-527-40604-2

Ferid Murad M
George E. Palade M
Arno A. Penzias P
Martin L. Perl P
William D. Phillips P
Norman F. Ramsey P
Robert Schrieffer P
William F. Sharpe E

Jack Steinberger P
Joseph H. Taylor Jr. P
Charles H. Townes P
Daniel C. Tsui P
Harold E. Varmus M
Robert W. Wilson P
Ahmed H. Zewail C

Appeal to Support an International Einstein Year

In the year 2005, scientists throughout the world will be celebrating the centenary of the theory of special relativity and the light-quantum hypothesis, both developed by Albert Einstein in 1905. The celebrations will also honour the 50th anniversary of Einstein's death in 1955.

Einstein was not only an extraordinary scientist, but also a scientist who faced his social responsibilities, intervened in political affairs and stood up and fought for civil rights. Throughout his life, he was committed to social justice, disarmament and peace.

As Einstein repudiated nationalistic attitudes and meaningless social rituals, the International Einstein Year 2005 should therefore reflect his universal and cosmopolitan stance.

The future of democratic societies rests on the comprehensive education and training of all its citizens. Scientific results must therefore be accessible to everyone. Education should not remain a privilege for the chosen few. The future of the citizens of all countries depends on the willingness of those who are prepared to commit themselves to a principle of solidarity whereby fair cultural and social services and economic trading, as well as an ecologically sound use of resources, are indispensable.

The future of humankind lies in the peaceful and tolerant cooperation between all countries and cultures. The elimination of atomic weapons and other means of mass destruction must therefore be the first and most important step in creating a world in which war as a means of solving conflicts no longer plays a role. To put it in Einstein's words: *"War cannot be humanized. It can only be abolished. "*

Scientists from all over the world are called upon to face up to their social responsibilities and to commit themselves to making scientific results the cultural heritage of all people. In doing so, poverty, under-development, and ecological destruction can be counteracted in

Einstein – Peace Now! Reiner Braun and David Krieger (Eds.)
Copyright © 2005 WILEY-VCH Verlag GmbH & Co. KGaA, Weinheim
ISBN 3-527-40604-2

a peaceful manner. In an interview in 1929, Einstein expressed his notion of a peaceful and commercially impartial world with the following words:

> "Think of what a world we could build if the power unleashed in war were applied to constructive tasks! One tenth of the energy that the various belligerents spent in the World War, a fraction of the money they exploded in hand grenades and poison gas would suffice to raise the standard of living in every country and avert the economic catastrophe of world wide unemployment. We must be prepared to make the same heroic sacrifices for the cause of peace that we make ungrudgingly for the cause of war. There is no task that is more important or closer to my heart. Nothing that I can do or say will change the structure of the universe. But maybe, by raising my voice, I can help the greatest of all causes – good will among men and peace on earth."

We, the initiators of this appeal in support of the International Einstein Year 2005, aim to realize this vision of the future in the spirit of the great scientist and call on all peoples of the world for their support.

The appeal is signed by:
Prof. Dr. Marian Ewurama Addy
University of Ghana, Ghana
Prof. Dr. Zhores I. Alferov
Nobel Prize Laureate in Physics 2000, Russia
Prof. Dr. Josef Altshuler
President of the Cuban Society for the History of Science and Technology, Cuba
Oscar Arias Sanchez
Former President of Costa Rica, Nobel Peace Prize Laureate 1987, Costa Rica
Prof. Dr. Abhay Ashtekar
Director of the Centre for Gravitational Physics and Geometry, India/USA
Prof. Dr. Fernando de Souza Barros
Pugwash Council, Brazil
Prof. Dr. Ulrike Beisiegel
Hospital of the University of Hamburg (UKE), Chairperson of the European Atherosclerosis Society, Germany

Prof Dr. Baruj Benacerraf
Nobel Prize Laureate in Medicine 1980, USA
Prof. Carlo Bernardini
Physicist at the First University of Rome, Italy
Prof. Dr. Hans Bethe
Nobel Prize Laureate in Physics 1967, USA
Prof. Dr. Andreas Buro
Peace researcher, Germany
Prof. Dr. Jeffrey Boutwell
Executive Officer at the American Academy of Arts and Science,
Executive Director of Pugwash Conferences on Science and World
Affairs, USA
Col. (ret.) Pierre Canonne
Former Head of TDB at the Organization for the Prohibition of
Chemical Weapons in The Hague and Pugwash Council, France
Ernesto Cardenal
Writer, Nicaragua
Prof. Dr. Arvid Carlsson
Nobel Prize Laureate in Medicine 2000, Sweden
Prof. Dr. Ana-Maria Cetto
Secretary-General of the International Council for Science, Mexico
Prof. Dr. Chen Ning Yang
Nobel Prize Laureate in Physics 1957, China
Prof. Dr. Paolo Cotta-Ramusino
Pugwash Secretary General, Italy
Prof. Dr. Paul Crutzen
Nobel Prize Laureate in Chemistry 1995, Germany/Netherlands
Prof. Nicola Cufaro Petroni
Secretary General of USPID, Union of Italian Scientists for
Disarmament
Prof. Dr. Jean Dausset
Nobel Prize Laureate in Medicine 1980, France
His Holiness the Dalai Lama
Nobel Peace-Prize 1989
Ambassador Jayantha Dhanapala
Former UN Under-Secretary General for Disarmament, Sri Lanka
Prof. Dr. Ogunlade Davidson
Scientific Advisor of UNEP, South Africa

Prof. Dr. Francisco Jose Delich
Former Rector of the National University of Cordoba and Buenos Aires, Argentina national Einstein Year
Prof. Dr. Johann Deisenhofer
Nobel Prize Laureate in Chemistry 1988, USA
Prof. Dr. Hans-Peter Dürr
MPI for Astrophysics, Alternative Nobel Prize Laureate 1987, Germany
Prof. Dr. Manfred Eigen
Nobel Prize Laureate in Chemistry 1967, Germany
Adolfo Perez Esquivel
Nobel Peace Prize Laureate 1980, Argentina
Prof. Richard R. Ernst
Nobel Prize Laureate in Chemistry 1991, Switzerland
Prof. Dr. Dietrich Fischer
Director of the European Peace Museum, USA/ Austria
Prof. Dr. John Kenneth Galbraith
Economist at Harvard University, USA
Prof. Dr. Johan Galtung
Alternative Nobel Laureate 1987, Norway
Prof. Dr. Vitaly Ginzburg
Nobel Prize Laureate in Physics 2003, Russia
Mikhail Gorbachev
Nobel Peace Prize Laureate 1990, Russia
Prof. Dr. Cive W.W.J. Granger
Nobel Prize Laureate in Economics 2003, USA
Prof. (Emeritus) Margherita Hack
Astonomical Observatory, Italy
Prof. Karen Hallberg
Pugwash Council, Board of Physics Association, Argentina
Prof. Dr. Gert Harigel
CERN, Switzerland
Prof. Dr. Herbert A. Hauptmann
Nobel Prize Laureate in Chemistry 1985, USA
Prof. Dudley R. Herschbach
Nobel Prize Laureate in Chemistry 1986, USA
Prof. Dr. Dieter B. Herrmann
Director of the Archenhold Observatory, Germany

Prof. Dr. Frank von Hippel
Princeton University, Former President of the
Federation of American Scientists (FAS), USA
Prof. Pervez Hoodbhoy
Pugwash Council, Pakistan
Prof. Dr. Roald Hoffmann
Nobel Prize Laureate in Chemistry 1981, USA
Prof. Dr. Tim Hunt
Nobel Prize Laureate in Physiology/Medicine 2001, UK
Prof. Dr. Eryk Infeld
Soltan Institute of Nuclear Studies, Poland
**International Physicians for the Prevention of Nuclear War
(IPPNW)**
Nobel Peace Prize Laureate 1985, USA
International Peace Bureau (IPB)
Nobel Peace Prize Laureate 1910, Switzerland
Prof. Dr. Gordana Jovanovic
University of Beograd, Serbia and Montenegro
Prof. Dr. Jerome Karle
Nobel Prize Laureate in Chemistry 1985, USA
Prof. Dr. Matthias Kreck
Professor of Mathematics, University of Heidelberg, Germany to
support an International Einstein Year
Prof. Dr. Walter Kohn
Nobel Prize Laureate in Chemistry 1998, USA
Prof. Dr. Masahashi Koshika
Nobel Prize Laureate in Physics 2002, Japan
Dr. David Krieger
President of the Nuclear Age Peace Foundation, USA
Vice President of the International Network of Engineers and
Scientists for global Responsibility (INES)
Prof. Dr. Paul Kurtz
President of the International Academy of Humanism, USA
Dr. David Lange
Former Prime Minister of New Zealand, Alternative Nobel Prize
Laureate 1983, New Zealand
Prof. Dr. Anne McLaren
Former Foreign Secretary of the Royal Society of theUnited
Kingdom, Pugwash Council, UK

Prof. Dr. Jean-Marie Lehn
Nobel Price Laureate in Chemistry 1987, France
Beisel Lemke
Alternative Nobel Prize Laureate 2000, Turkey
Prof. Dr. Rita Levi-Montalcini
Nobel Prize Laureate in Medicine 1986, Italy
Prof. Dr. Jiri Matousek
Advisor to the OPCW, University of Brno, Czech Republic
Prof. Dr. Claus Montonen
President of the International Network of Engineers and Scientists
for global Responsibility (INES), Finland
Prof. Dr. Phil Morrison
(MIT) Manhattan-Project, Founder of the Federation of American
Scientists (FAS), USA
Robert O. Muller, Thomas Gebauer
Co-founder of the "International Campaign to Ban Landmines",
Nobel Peace Prize Laureate 1997, USA/Germany
Prof. Dr. Erwin Neher
Nobel Prize Laureate in Medicine/ Physiology 1991, Germany
Dr. Götz Neuneck
Pugwash Council, Germany
Prof. Dr. Kathryn Nixdorff
Interdisziplinäre Arbeitsgruppe Naturwissenschaft, Technik
und Sicherheit (IANUS), Germany
Prof. Dr. Hitoshi Ohnishi
Pugwash Council, Vice President of Tohoku University, Japan
Prof. Dr. Luis de la Pena Averbach
Former President of the Mexican National
Society of Physics, Mexico
Prof. Dr. Valery Petrosjan
Director of the Department of Chemistry, Lomonosow University,
Russia
Prof. Dr. Hugo Perez
Director at the Institute of Cybernetic Mathematics and Physics,
Cuba
Prof. Dr. Peter H. Plesch
Professor of Chemistry, University of Keele, UK

Prof. Dr. Jürgen Renn
Director at the Max-Planck-Institute for the History of Science, Germany
Prof. Dr. Horst–Eberhard Richter
Director of the Sigmund-Freud-Institute, Germany
Senator Douglas Roche
President of the Middle Power Initiative (MPI), Canada
Prof. Dr. Joseph Rotblat
Nobel Peace Prize Laureate 1995, UK
Acad. Dr. Yury Ryzhov
Pugwash Council, former member of the Presidential Council, Russia
Arundhati Roy
Writer, India
General (ret.) Mohammed Kadry Sahid
Al-Ahram Centre for Political and Strategic Studies, Pugwash Council, Egypt
Prof. Dr. Frederick Sanger
Nobel Prize Laureate in Chemistry 1958 and 1980, UK
Dr. Hermann Scheer
Alternative Nobel Prize Laureate 1999, Germany
Prof. Dr. Jürgen Schneider
Board member of German Initiative Science for Peace and Sustainability
Mycle Schneider
Alternative Nobel Prize Laureate 1997, France
Prof. Dr. John Stachel
Director of the Boston University Centre for Einstein Studies, USA
Prof. Dr. Dhirendra Sharma
Director, Centre for Science Policy Research, India
Prof. Dr. Jack Steinberger
Nobel Prize Laureate in Physics 1988, CERN, Switzerland
Dr. Marc Byung-Moon Suh
Pugwash Council, South Korea
Prof. Dr. Igor Tipans
Technical University Riga, Latvia
Archbishop Desmond Tutu
Nobel Peace Prize Laureate 1984, South Africa

Prof. Dr. Joseph H. Taylor
Nobel Prize Laureate in Physics 1993, USA
Prof. Dr. John Walker
Nobel Prize Laureate in Chemistry 1997, UK
Dr. Jakob von Uexküll
Founder of the Alternative Nobel Prize/ Right
Livelihood Award, Sweden/ Germany
Dr. Paul Walker
International Global Green Cross, USA
Dr. Phil Webber
Chairman of Scientists for Global Responsibility, (SGR), UK
Prof. Dr. Joseph Weizenbaum
Former Massachusetts Institute of Technology, USA/ Germany
Prof. Dr. Carl-Friedrich von Weizsäcker
Germany
Prof. Dr. Manfred Wekwerth
Stage director, Germany
Harry Wu
China/ USA
Prof. Dr. Herbert Wulf
Former Director of the Bonn International Conversion Centre,
Germany
Dr. Alla Yaroshinskaja
Alternative Nobel Prize Laureate 1992, Russia

International Network of
Engineers and Scientists for
global Responsibility

Berlin Information-Centre for
Transatlantic Security

Editors: Reiner Braun, Dipl.-Ing. Nicola Hellmich, Ekkehard Sieker

With the support of Global Green-Green Cross, Germany

Acknowledgements

I would like to thank all authors who have taken the time, beside their numerous other commitments, to support this project by contributing their thoughtful and inspiring articles to the book.

Thanks are also due to the Wiley-VCH Publishing House, most of all to Dr. Alexander Grossmann, for his cooperation and management of the whole project, but also to Esther Dörring and Anja Tschörtner, who dealt with the manuscript and handled the editorial part of the project.

I am grateful to the Max-Planck-Institute for the History of Science, especially to its director Jürgen Renn, who enabled me to co-edit this book in the first place.

I am also indebted to my friends Peter Badge, Ursula Schefler and Ekkehard Sieker, for supporting and helping me with this project, the publication of which would not have been possible without them. I would particularly like to thank Nicola Hellmich, whose collaboration in advancing the project was an enormous help throughout.

Finally I would like to express my gratitude to Melzer Verlag, where the German edition of this book is published.

Einstein – Peace Now! Reiner Braun and David Krieger (Eds.)
Copyright © 2005 WILEY-VCH Verlag GmbH & Co. KGaA, Weinheim
ISBN 3-527-40604-2